Carmen Campanile

Imaging techniques and therapeutic tools for osteosarcoma

Carmen Campanile

Imaging techniques and therapeutic tools for osteosarcoma

Osteosarcoma: imaging and treatment

Südwestdeutscher Verlag für Hochschulschriften

Impressum / Imprint

Bibliografische Information der Deutschen Nationalbibliothek: Die Deutsche Nationalbibliothek verzeichnet diese Publikation in der Deutschen Nationalbibliografie; detaillierte bibliografische Daten sind im Internet über http://dnb.d-nb.de abrufbar.

Alle in diesem Buch genannten Marken und Produktnamen unterliegen warenzeichen-, marken- oder patentrechtlichem Schutz bzw. sind Warenzeichen oder eingetragene Warenzeichen der jeweiligen Inhaber. Die Wiedergabe von Marken, Produktnamen, Gebrauchsnamen, Handelsnamen, Warenbezeichnungen u.s.w. in diesem Werk berechtigt auch ohne besondere Kennzeichnung nicht zu der Annahme, dass solche Namen im Sinne der Warenzeichen- und Markenschutzgesetzgebung als frei zu betrachten wären und daher von jedermann benutzt werden dürften.

Bibliographic information published by the Deutsche Nationalbibliothek: The Deutsche Nationalbibliothek lists this publication in the Deutsche Nationalbibliografie; detailed bibliographic data are available in the Internet at http://dnb.d-nb.de.

Any brand names and product names mentioned in this book are subject to trademark, brand or patent protection and are trademarks or registered trademarks of their respective holders. The use of brand names, product names, common names, trade names, product descriptions etc. even without a particular marking in this works is in no way to be construed to mean that such names may be regarded as unrestricted in respect of trademark and brand protection legislation and could thus be used by anyone.

Coverbild / Cover image: www.ingimage.com

Verlag / Publisher:
Südwestdeutscher Verlag für Hochschulschriften
ist ein Imprint der / is a trademark of
AV Akademikerverlag GmbH & Co. KG
Heinrich-Böcking-Str. 6-8, 66121 Saarbrücken, Deutschland / Germany
Email: info@svh-verlag.de

Herstellung: siehe letzte Seite /
Printed at: see last page
ISBN: 978-3-8381-3007-1

Zugl. / Approved by: Zürich, ETH, Diss., 2013

Copyright © 2013 AV Akademikerverlag GmbH & Co. KG
Alle Rechte vorbehalten. / All rights reserved. Saarbrücken 2013

There is something fascinating about science.
One gets such wholesale returns of conjecture
out of such a trifling investment of fact.

TABLE OF CONTENTS

SUMMARY	6
LIST OF ABBREVIATIONS	9
INTRODUCTION: OSTEOSARCOMA	11
1.1 INCIDENCE AND ETIOLOGY	12
1.2 METASTASES	14
1.2.1 METASTASES IN OSTEOSARCOMA	18
1.3 DIAGNOSIS	20
1.3.1 CLINICAL PRESENTATION	20
1.3.2 DIAGNOSTIC IMAGING TOOLS	20
1.3.3 BIOPSY	23
1.3.4 HISTOLOGY	24
1.3.5 STAGING	29
1.4 PROGNOSIS AND TREATMENT	30
1.4.1 CLINICAL FACTORS	31
1.4.2 TREATMENT	34
1.4.3 CHEMORESISTANCE IN OS	38
CHAPTER 2	40
INTRODUCTION: POSITRON EMISSION TOMOGRAPHY AND PHOTODYNAMIC THERAPY	40
2.1 POSITRON EMISSION TOMOGRAPHY	41
2.1.1 PET IN THE CLINICS	43
2.1.2 PET IN ONCOLOGY	46
2.2 PHOTODYNAMIC THERAPY	57
2.2.1 BASIC PRINCIPLE	57
2.2.2 PHOTOSENSITIZERS	58
2.2.3 MECHANISMS OF PDT ACTION ON THE TUMOUR	61
2.2.4 PDT IN THE CLINICS	69
2.2.5 PDT IN OSTEOSARCOMA	72
2.2.6 APPLICATION ROUTES FOR PDT	73
2.3 AIMS OF THE THESIS:	75
2.3.1 OS MOUSE MODELS	76
CHAPTER 3	83

EVALUATION OF PET IMAGING IN OS MOUSE MODELS 83

3.1 RESULTS — 84
3.1.1 UPTAKE OF ^{18}F-FET, ^{18}F-FLT AND ^{18}F-FCH IN THE 143B MODEL — 84
3.1.2 ^{18}F-FDG UPTAKE IN THE THREE OS MOUSE MODELS — 86
3.1.3 ^{18}F-FMISO UPTAKE IN THE THREE OS MOUSE MODELS — 87
3.1.4 ^{18}F-FLUORIDE UPTAKE IN THE THREE OS MOUSE MODELS — 88
3.1.5 IMMUNOHISTOCHEMISTRY OF OS MODELS USED IN THIS STUDY — 89
3.1.6 DETECTION OF METASTASES VIA PET IMAGING — 93
3.2 DISCUSSION & OUTLOOK — 96
3.3 MATERIAL & METHODS — 105
3.3.1 CELL LINES — 105
3.3.2 MOUSE MODELS — 105
3.3.3 PRIMARY TUMOUR VOLUME — 106
3.3.4 RADIOTRACER SYNTHESIS — 106
3.3.5 PET SCANS — 106
3.3.6 X-Gal STAINING — 108
3.3.7 QUANTITATIVE ANALYSIS AND CALCULATION OF THE SENSITIVITY — 109
3.3.8 HISTOLOGY AND IMMUNOHISTOCHEMISTRY — 110
3.3.9 Ki67 AND CaIX INDEX — 110
3.3.10 VESSEL AREA — 111
3.3.11 AUTORADIOGRAPHY OF THE LUNGS — 112
3.3.12 STATISTICS — 112

CHAPTER 4 113

EVALUATION OF EFFICACY OF PHOTODYNAMIC THERAPY IN OS 113

4.1 RESULTS — 114
4.1.1 HIGHER PS UPTAKE IN THE HIGHLY METASTATIC OS CELL — 114
4.1.2 143B CELLS ARE HIGHLY SENSITIVE TO THE PDT TREATMENT — 117
4.1.3 APOPTOSIS -INDUCED PHOTO-TOXICITY — 120
4.1.4 HIGHLY SELECTIVITY OF PS UPTAKE *IN VIVO* — 121
4.1.5 OPTIMISAZION OF PDT PROTOCOLS FOR OS TREATMENT IN MICE — 128
4.2 DISCUSSION & OUTLOOK — 136
4.3 MATERIAL & METHODS — 142
4.3.1 MATERIAL AND OS CELL LINES — 142
4.3.2 UPTAKE OF LIPOSOMAL FORMULATION 1 — 143
4.3.3 CYTOTOXICITY ASSAY — 144
4.3.4 WESTERN BLOT ANALYSIS — 145
4.3.5 MOUSE MODELS — 146
4.3.6 *IN VIVO* UPTAKE OF LIPOSOMAL FORMULATION 2 — 147
4.3.7 TREATMENT PROTOCOL — 148

CHAPTER 5 — 150

CONCLUSIONS	151
6 REFERENCES	154
7 ACKNOWLEDGMENTS	164

SUMMARY

Osteosarcoma (OS) is the most common malignant bone tumour in children and young adolescents. Many efforts have been made during the past decades to improve the survival of OS patients and indeed the combination of resection of the primary tumour and neo-adjuvant chemotherapy increased the 5-year survival of patients with a localised disease from 20 to 80%. However patients with metastatic disease still have low chances to survive: the 5-year survival remained at 20% since the second half of the 1970s.

Consequently, clinicians who take care of OS patients face two important challenges: first they need imaging tools with much higher sensitivity and resolution than current equipment for early detection of primary tumours and metastases and better monitoring of the therapy response. Second, tumour staging needs to be improved for the design of more tumour-specific treatment strategies and the development of new drugs that more selectively target and kill tumour cells.

Therefore, in this thesis, we aimed at assessing on one side the potential use of positron emission tomography (PET) as an additional tool *in vivo* for the characterisation of different OS phenotypes and for the detection of lung metastases. On the other side we evaluated the efficacy of photodynamic therapy (PDT) for the treatment of OS.

For the first aim we validated the power of PET imaging making use of two xenograft, 143B and SaOS-2/Caprin-1, and one syngeneic, LM8, orthotopic OS mouse model that represent the heterogeneity of OS phenotypes in the human disease, where the tumour cells metastasise towards the lung. We analysed, at the beginning, the uptake of six PET tracers in the primary tumour and compared it to the uptake in the control healthy leg. The tracers used were Fluoro-2-deoxy-D-glucose (^{18}F-FDG), an indicator of glucose metabolism,

Fluormisonidazole (^{18}F-FMISO), an indicator of hypoxia, ^{18}F-Fluoride, an indicator of bone remodeling, 3'-Fluoro-3'-deoxythymidine (^{18}F-FLT), an indicator of DNA synthesis, O-(2-Fluoroethyl)-L-tyrosine (^{18}F-FET), an indicator of protein synthesis and Fluoromethylcholine (^{18}F-FCH), indicator of membrane turnover. The last three tracers did not show any consistent results in one representative xenograft model, namely 143B. The uptake of the other three tracers (^{18}F-FDG, ^{18}F-FMISO and ^{18}F-Fluoride) was, instead, evaluated in all three tumour models and we found that PET can provide additional information on the biology of individual tumours and on predominant processes affecting the tumour development, thereby helping in selecting a better therapeutic strategy without neglecting the unequivocal value of a tumour biopsy.

In the predominantly osteolytic model, namely 143B, we observed a significant 3.7 fold increase of ^{18}F-FDG uptake indicating high glucose metabolism. In addition we recognized areas of hypoxia and found a significant 2 fold increase of ^{18}F-FMISO uptake in the primary tumour leg compared to the healthy control leg; bone remodelling was either absent or moderate (1.1 fold increase of ^{18}F-Fluoride uptake in the tumour leg). In the mouse model with only mild osteoblastic lesions, on the other hand, a significant but more modest uptake of glucose (2.2 fold increase) compared to the osteolytic model was observed. Hypoxia was minimal and moderate bone remodelling with a 1.3 fold increase in ^{18}F-Fluoride uptake in the tumour leg. Finally, the mouse model with more pronounced osteoblastic lesions showed a 2 fold increase of glucose metabolism in the primary tumour leg, large areas of hypoxia and bone remodelling. Indeed we could observe a 2.2 fold higher uptake of ^{18}F-FMISO and 2.3 fold higher uptake of ^{18}F-Fluoride in the primary tumour compared to the control leg. Immunohistochemical staining of tumour tissue confirmed the results obtained with the different PET tracers.

Regarding the metastases, we could not detect foci in the lung with PET imaging in the SaOS-2/Caprin-1 and in the 143B mouse models. However we could confirm the presence of metastases *ex vivo* making use of X-Gal staining since the tumour cells were *LacZ* tagged.

For the second aim we focused on the development of new modalities for the treatment of OS, evaluating the efficacy of photodynamic therapy (PDT) in OS cell lines *in vitro* and optimising protocols for PDT *in vivo* in orthotopic OS mouse models.

In vitro we demonstrated that the highly metastatic 143B OS cells took up 2.5 fold more photosensitizer (PS) than the parental low metastatic HOS cells. Furthermore photo-toxicity was observed in the 143B cells already at a concentration of PS as low as 0.075 μgml^{-1} and was found mediated by apoptotic mechanisms leading to cell death, which occurred already 90 min after illumination of PS accumulating cells.

In view of these promising results, we focused on a well-established orthotopic OS mouse model generated by intratibial injection of 143B cells. We were able to define the concentration range of PS tolerated and being effective in our model (between 0.05 $mgkg^{-1}$ and 0.1 $mgkg^{-1}$).

In conclusion, the combination of improved imaging tools and novel treatment modalities such as PET and PDT, respectively, will in the future help to more accurately diagnose tumour phenotypes and improve treatment efficacy.

LIST OF ABBREVIATIONS

2-DG	2-Deoxyglucose
ALA	Aminolevulinic acid
ALP	Alkalyne phosphatase
BCG	Bacille- Colmette Guérin
CLSM	Confocal laser scanning microscopy
CT	Computed Tomography
CP	Corynebacterium parvum
Cu-ATSM	Cu(II)- diacetyl-bis(N4 methylthiosemicarbazone
DCA	Dichloroacetate
EDTA	Ethylenediaminetetraacetic acid
EGFP	Enhanced green fluorescent protein
ET_AR	ET-1 receptor
FCH	Fluoromethylcholine
FDG	Fluoro-2-deoxy-D-glucose
FET	O-(2-Fluoroethyl)- L-tyrosine
FLT	3'-Fluoro-3'-deoxythymidine
FMISO	Fluoromisonidazole
GLUT	Glucose Transporter
HD-MTX	High-dose methotrexate
H&E	Hematoxylin and Eosin
HIF 1 α	Hypoxia-inducible factor 1 α
HPD	Hematoporphyrin derivative
HSV1-tk	Herpes simplex virus type 1 thymidine kinase
IHC	Immunohistochemistry
IL-6	Interleukin-6
INF α or γ	Interferon α or γ
LDH	Lactate dehydrogenase
MBq or Bq	Megabequerel or bequerel
MRI	Magnetic Resonance Imaging
mTHPC	5,10,15,20-tetrakis(meta-hydroxyphenyl)chlorine
OS	Osteosarcoma
PARP	Poly-ADP-ribose polymerase
PDT	Photodynamic Therapy
PEDF	Pigment epithelium-derived factor
PET	Positron Emission Tomography
PFA	Paraformaldehyde
PS	Photosensitizer
RFU	Relative fluorescent unit

ROI	Region of interest
SEM	Standard error of the mean
SPECT	Single-photon emission computed tomography
SUV	Standardised uptake value
TCA	Tricarboxylic acid cycle
TLRs	Toll-like receptors
TNF	Tumour necrosis factor
TPZ	N-oxyde tirapazamine
ZOL	Zolendronic acid

CHAPTER 1

INTRODUCTION: OSTEOSARCOMA

1.1 INCIDENCE AND ETIOLOGY

Osteosarcoma (OS) is a rare malignant primary bone tumour accounting for 0.2% of all the human tumours but it is the most common cancer of the bone (35%) followed by chondrosarcoma (25%) and Ewing sarcoma (16%) in children and young adults (Fletcher 2002; Mirabello, Troisi et al. 2009). Indeed the incidence reaches 11/1.000.000/year in adolescents between 15 and 19 years of age while the overall incidence is attested on 3/1.000.000/year (Hogendoorn, Athanasou et al. 2010). Despite the high frequency in the young population it shows a second peak of incidence in the seventh and eight decade of life (Fig.1.1). In the elderly patients OS is usually related to previous radiotherapy for other cancers' treatment or secondary to pathologies like Paget's disease of bone, hereditary retinoblastoma, Li-Fraumeni-, Werner-, Rothmund Thomson- and Bloom-syndromes (Price 1955; Huvos 1986; Hogendoorn, Athanasou et al. 2010).

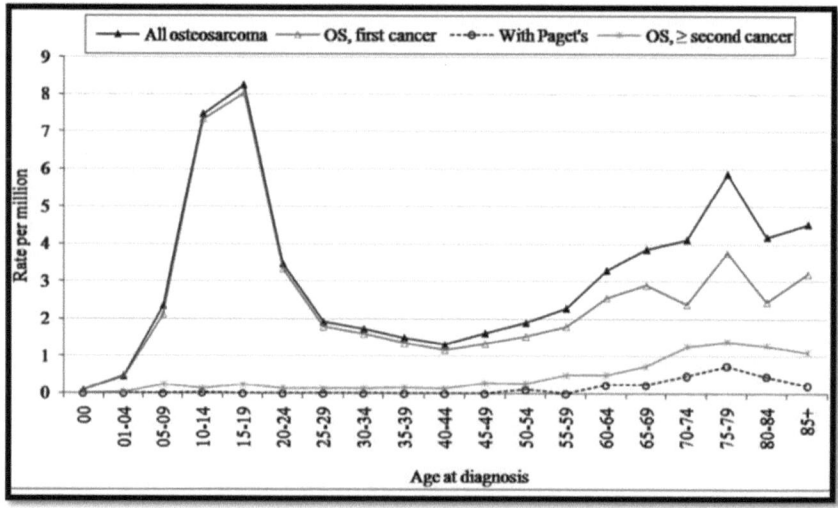

Fig.1.1 Osteosarcoma age-related incidence according to the Surveillance, Epidemiology and End Results (SEER) program (Mirabello et al., *Cancer* 2009)

These pathologies are usually associated with hereditary modifications that are listed in the Table 1.1 below and in the case of OS they show a complex karyotype with chromosomal rearrangements that affect mainly the chromosome stability and the cell cycle regulation but that are not recurrent in all the OS patients (Helman and Meltzer 2003).

Disorder	Gene	Protein
Li-Fraumeni Syndrome	TP53	p53
Retinoblastoma	RB1	Retinoblastoma-1
Rothmund Thomson Syndrome	REQL4	RecQ protein-like 4, DNA helicase
Werner Syndrome	WRN	RecQ helicase like
Bloom Syndrome	BLM	RecQ helicase like
Diamond Blackfan Anemia	RPS19, RPL5, RPL11	Ribosomal proteins

Table 1.1: Genetic disorders and their most frequent gene mutation found in elder patients with OS

The most common genetic alterations found in conventional OS include loss of heterozygosity of chromosome regions in 3q, 13q, 17p, 18q and amplification in 1q21-23 and 17p (Dorfman 1998).

In young patients the aetiology is unknown but it is most frequently related to the rapid growth of the bones therefore it is most frequent at the metaphyses of the long bones where there is high grade of cell proliferation and turnover during adolescence. Indeed it was shown that OS is more common in tall people compared to short ones which relates with the previous statement concerning the pubertal growth (Klein and Siegal 2006).

In the elder people the lower long bones are still the most common OS location but to a less extent.

In both groups of age the overall rate of frequency is 42-45% in the femur, 19% in the tibia, 10-11% in the humerus, 8% in the skull or jaw and 8% in the pelvis (Ottaviani and Jaffe 2009; Abed and Grimer 2010). Male show higher rate of OS incidence than female (1.35 : 1) but it is described that female younger than 15 years are slightly more affected than men as well as elder black women with a previous diagnosed cancer (Mirabello, Troisi et al. 2009; Ottaviani and Jaffe 2009; Savage and Mirabello 2011). In the table 1.2 you see the results from a study conducted in Europe where they describe the higher incidence of OS in the girls younger than 15 and in the boys above 15 years (Stiller, Bielack et al. 2006).

	Boys		Girls	
Europe	0-14 years	15-19 years	0-14 years	15-19 years
British Isles	2.6	9.7	2.6	4.7
East	2.3	8.1	2.7	3.3
North	3.0	11.6	2.7	5.8
South	3.1	12.2	4.2	4.3
West	2.9	10.5	2.7	7.8
Total	2.78	10.42	2.98	5.18

Table 1.2: Summary of annual incidence rates per million for OS between 1988 and 1997 in the five areas of Europe (adapted from (Stiller, Bielack et al. 2006)).

1.2 METASTASES

Metastases in OS still remain the main cause of death in patients, and more specifically, 90% of cancer related deaths are due to metastatic spread in different organs mainly because the metastatic cells are the most resistant to therapies (Mehlen and Puisieux 2006; Nguyen and Massague 2007; Mathot and Stenninger 2011).

This multi-step event starts with the "pre-metastatic" cells inducing angiogenesis in situ, followed by the escape from the original primary tumour mass. After adhering and entering the blood vasculature either directly through blood vessels or passing, first, through the lymphatic system, these cells migrate to the target organ, they extravasate and colonise the target organ (Fig.1.2).

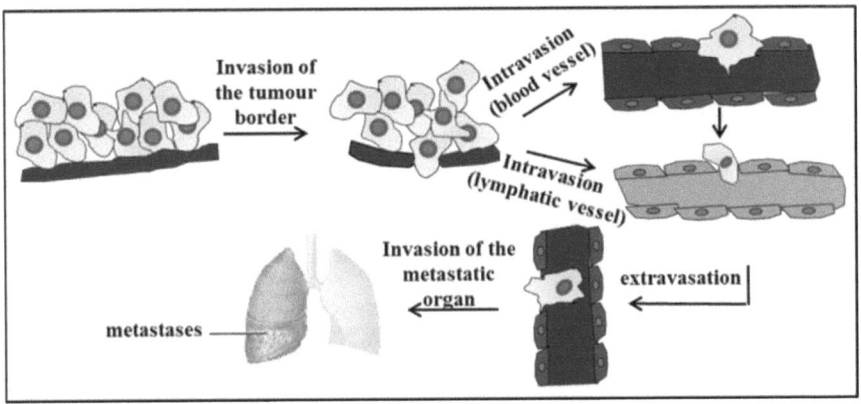

Fig.1.2 Schematical view of the metastatic process [adapted from (Steeg 2003)]

According to the "seed and soil" theory, the migration is not casual but "pre-metastatic" cells rather prefer to migrate toward specific organs (see table 1.3) (Dai, Haqq et al. 2006; Nguyen, Bos et al. 2009). Indeed Stephen Paget in 1889 launched the "seed and soil" theory after having realised that most of his patients with breast cancer showed metastases only in the liver and not in the spleen while patients with sepsis died with abscesses in both organs. Therefore he thought that primary tumour cells behave like seeds that would migrate, grow and survive only where they find the appropriate ground (soil).

More than 30 years later James Ewing came up with another theory: it is a combination of mechanical forces and circulatory flow that gives organ-specificity for the metastases formation. This can explain the migration towards lung and liver but in the bone the circulation cannot explain the high rate of metastases: in this case the presence of the bone microenvironment, consisting of bone stromal cells, hematopoietic elements, adipocytes and the high expression of growth and regulatory factors makes the skeleton a comfortable "soil" for tumour cells (Dai, Haqq et al. 2006).

Tumour Type	Principal organs of metastases
Breast	bone, lungs, liver and brain
Lung adenocarcinoma	brain, bones, adrenal gland and liver
Skin melanoma	lungs, brain, skin and liver
Colorectal	liver and lungs
Pancreatic	liver and lungs
Glioblastoma	Central nervous system (no distant metastases)
Neuroblastoma	bone, liver and lung
Prostate	bones
Myeloma	bones
Sarcoma	lungs

Table 1.3: Main location of metastases for some primary tumours (adapted from table 1 in (Nguyen and Massague 2007; Nguyen, Bos et al. 2009)).

What still remains to be fully understood is the latency of some tumours in metastasising: lung and breast adenocarcinomas metastasise both to the same organs (bone, liver and brain) but while the lung cancer metastasises usually within few months, in the breast cancer the metastases develop after many years (Mathot and Stenninger 2011). One explanation for this phenomenon is given by Nowell that speaks about the need for the tumour cells to acquire many mutations in order to gain adaptability and consequently metastatic potential and this hypothesis might explain why some tumours need longer time to metastasise compared to other tumours (Nowell 1976). Furthermore another explanation comes from recent studies that highlight also the importance of a specific microenvironment made of host cells that actively participate in the tumour expansion and growth towards other organs (Mareel, Oliveira et al. 2009; Hanahan and Weinberg 2011).

Different are the classes of genes involved in the different steps of the metastatic process and Nguyen et al. identifies three classes: metastasis initiation, progression and virulence genes (Nguyen, Bos et al. 2009).

The initiation genes allow the tumour cells to gain the features for a more aggressive phenotype like TWIST, SNAI1 and SNAI2 that induce the epithelial-mesenchymal transition or metadherin in breast cancer and the metastasis-associated in colon cancer 1 (MACC1) in colorectal carcinoma that contribute to the invasive phenotype of the "pre-metastatic" cells (Yang and Weinberg 2008; Hu, Chong et al. 2009; Stein, Walther et al. 2009).

The progression genes are involved in providing the "pre-metastatic cells" with the right features to cross the endothelial cells, migrate in the blood vessel and settle in the parenchima of the new organ (Nguyen, Bos et al. 2009). CXCR4-CXCL12 axis is the most studied and understood chemokine pair in the progression of cancer: in human prostate cancer specimens CXCR4 expression correlated with the aggressiveness of the tumour as well as in breast cancer (Kato, Kitayama et al. 2003; Sun, Wang et al. 2003). Adhesion molecules, integrins, selectins belong to this class of genes, since they contribute in the different steps of the intravasation, cell migration and extravasion (Dai, Haqq et al. 2006). Depending on which genes are de-regulated, tumour cells and metastases will migrate and invade different organs.

Finally there are the metastasis virulence genes whose deregulation conveys a selective genes expression for metastatic cells in the specific organ: metastatic cells from a breast primary tumour can, at a later time point, express parathyroid hormone-related protein and interleukin-11 which leads to the formation of osteolytic metastases in bone (Yin, Selander et al. 1999; Mundy 2002; Kang, Siegel et al. 2003; Nguyen, Bos et al. 2009).

1.2.1 METASTASES IN OSTEOSARCOMA

In OS metastases occur preferentially in the lung (80-90%) and to a less extent in the bone (10-20%) and in the lymph nodes (9%) (Kaste, Pratt et al. 1999; Bielack, Kempf-Bielack et al. 2002; Longhi, Errani et al. 2006; Ritter and Bielack 2010).

Lung is frequently invaded by metastatic cells in different cancers and this is due first of all to the physiological high blood flow in this organ. Recent studies highlight the most important mediators in the pulmonary metastases coming from a primary breast cancer: epiregulin, prostaglandine G/H synthase 2, matrix metallo-proteinase 1 and 2 which act on the vasculature remodelling but also on the extravasion towards the lung and finally cytokine angiopoietin-like 4 that helps tumour cells to enter the lung by destroying the endothelial cells interaction (Gupta, Nguyen et al. 2007; Padua, Zhang et al. 2008; Nguyen, Bos et al. 2009).

In OS proteins involved in metastatic progression are: von Willebrand factor, whose expression was found in tumour cells passing to a metastatic phenotype, is a glycoprotein that is involved in the platelet aggregation at the subendothelial matrix and might therefore help the tumour mass to be hidden among platelets in the blood vessels and escape the surveillance from the immune system (Eppert, Wunder et al. 2005).

An important pair involved in the selective migration of OS cells towards the lung is CXCR4-CXCL12 whose contribution in metastatic progression was discussed in the previous paragraph in the prostate and breast cancer. In OS, CXCR4 expressing tumour cells follow a gradient of CXCL12 that is abundant in the lung: treatment with CXCR4 inhibitors or antagonists of

OS bearing mice showed a decreased amount of metastases in the lung (Perissinotto, Cavalloni et al. 2005; Kim, Lee et al. 2008).

Finally the receptor activator nuclear factor κB ligand (RANKL) was found highly expressed in bone metastatic cells of breast and prostate cancer where it is responsible for increased cell migration and metastatic proneness: in OS it correlates with a poor survival in patients with high-grade localised OS proposing a cause-effect link between RANKL and metastases which still needs to be confirmed (Lee, Jung et al. 2011).

1.3 DIAGNOSIS

1.3.1 CLINICAL PRESENTATION

Diagnosis of OS is rather difficult and it is mandatory to have a multidisciplinary group that can interpret the results and plan proper strategies of treatment and follow up.

The most common clinical phenomenon is localised pain, usually in the night and not connected to any specific activity, and which can last up to several months (Abed and Grimer 2010; Ritter and Bielack 2010). In 1/3 of the cases, a tumour mass is visible at the first visit already which is usually caused by the soft tissue swelling but only few patients complain about it (Widhe and Widhe 2000; Picci 2007). When the tumour is osteolytic, the patient may present with a pathological fracture (10-15% of the cases) and if this is present then the survival rate is lower because of high risk of recurrence (Scully, Ghert et al. 2002; Picci 2007; Ritter and Bielack 2010). Loss of appetite, weight loss, fever and pallor are only rarely noticed. Finally laboratory analyses on blood samples are not specific for OS: lactate dehydrogenase (LDH) and alkaline phosphatases (ALP) are found elevated and they might be helpful for the final diagnosis (Picci 2007; Heare, Hensley et al. 2009; Biermann, Adkins et al. 2010). In this concern Bacci et al. noticed a correlation between high ALP and risk of relapse as well as high level of LDH probably might be associated with a worse prognosis but none of these markers is so far considered an OS marker (Meyers, Heller et al. 1992; Bacci, Picci et al. 1993; Bacci, Ferrari et al. 1994).

1.3.2 DIAGNOSTIC IMAGING TOOLS

After the clinical assessment, the next step is to perform radiography of the tumour that is followed by histological analysis of the biopsy in case of suspicious lesions. The radiograph can show osteoblastic, osteolytic or mixed lesions with irregular areas of calcifications in the soft tissue. A characteristic OS element is called "Codman triangle", because the calcified areas in the periosteal region form a triangle between the healthy tissue and the tumour cells (Ritter and Bielack 2010) (Fig.1.3).

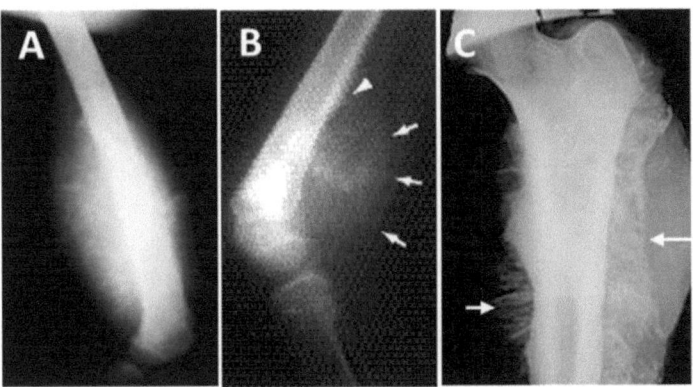

Fig.1.3: Lateral x-ray picture of OS in the distal femur with a classical dense tumour mass with patchy filaments that develop like a sunburst reaction (A); X-ray picture of OS in the distal femur with the characteristic Coldman triangle shown by the arrow head and with an extended soft tissue mass depicted by the arrows (B); X-ray picture of OS in the tibia with two typical imaging effects from the tumour namely end-hair (small arrow) and onion skinning (big arrow) (Dorfman 1998; Klein and Siegal 2006).

Together with radiography usually the Magnetic Resonance Imaging (MRI) is used to evaluate the extension of soft tissue involvement and the invasion of the tumour cells into the bone marrow (Sundaram 1997; Brenner, Bohuslavizki et al. 2003; Ritter and Bielack 2010). MRI and color-coded duplex sonography are included in the diagnostic workup when information about vasculature involvement is explored (Brenner, Bohuslavizki et al. 2003). Bone scintigraphy is also performed to define the border of the primary tumours (Picci 2007). Finally a chest Computed Tomography (CT) and a bone scintigraphy are performed to detect, respectively, lung or bone metastases and consequently define the tumour staging (Brenner,

Bohuslavizki et al. 2003; Abed and Grimer 2010). Around 15-25% of the patients show metastases at the time of diagnosis but it is believed that many more patients (around 80-90%) have at this time metastases but they are not detectable (Friedman and Carter 1972; Meyer 1991; Brenner, Bohuslavizki et al. 2003).

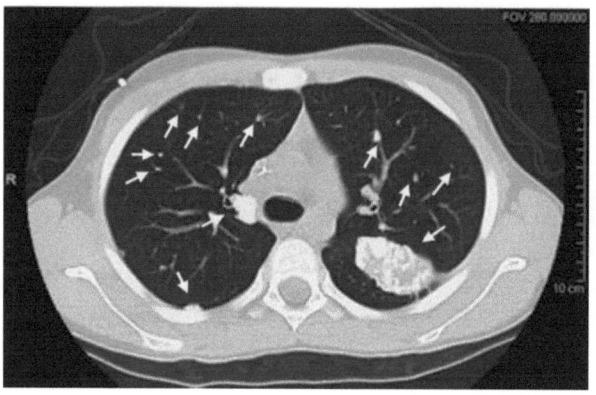

Fig.1.4: Computed Tomography (CT) of the lung showing presence of lung metastases (arrows) (Bacci, Rocca et al. 2008)

The chest CT is preferred to the chest radiography because of higher sensitivity that reaches 2mm for the diameter of detectable nodules but both methodologies do not give specific information about the nature of the nodules (Fig.1.4) (Davis 1991). A study conducted on 20 tumours including lymphomas, neuroblastomas, renal tumours, hepatic tumours and malignant bone tumours states that most of the lung nodules are < 5mm which would not have been visible on the radiography and the nodules > 5mm are usually malignant while nodules < 5mm were likely to be malignant in 1 case out of 9 benign nodules (Silva, Amaral et al. 2010).

These findings are not in agreement with another study where McCarville et al. claim that the lung nodules < 5mm were likely to be malignant or benign with the same probability but they analysed a lower number of scans compared to the previous study. In addition McCarville et al. believes that if the lung nodules diameter does not increase in 2 years, the probability is higher that it is only a benign nodule. Finally it is noticeable that there are some differences between adults and children in the diagnosis of lung nodules: in adults the irregular nodules are most probably malignant while in children the clearly defined ones are malignant as well as the smaller nodules (< 5mm) are more probably malignant (Bateson 1965; Ost, Fein et al. 2003; Iwano, Makino et al. 2004; Li, Sone et al. 2004; Ost and Fein 2004; McCarville, Lederman et al. 2006).

In summary CT is still the gold standard for lung nodules detection among the available new imaging tools, but the main disadvantage of CT is the difficulty in distinguishing between benign and malignant tumours. Therefore new interest is given to Positron Emission Tomography (PET) in combination with CT in the identification of malignant lung nodules since both techniques are complementary to each other. Indeed a study shows the possibility to distinguish between benign and malignant pulmonary lesions in patients with suspected primary or recurrent lung cancer using ^{18}F-Fluorodeoxyglucose (^{18}F-FDG) PET imaging: a standardised uptake ratio (SUR) was calculated for each patient on a region of interest (ROI) of 1cm in diameter and it was found that 2.5 was the best cut-off value to visualise malignant lesions with 97% sensitivity, 82% specificity. When the SUR was < 2.5, the benign lesions could be detected with a sensitivity of 82%, specificity of 97% (Duhaylongsod, Lowe et al. 1995).

1.3.3 BIOPSY

Biopsy is the gold standard for the final diagnosis. An experienced orthopaedic oncologist performs the invasive procedure keeping in mind that the tumour piece should represent the tumour material and contain no normal tissue; part of the tumour tissue should be used for cell culture and part for the histology and in the tumour centres a small piece should be frozen immediately after the biopsy for further analysis and a request form should be filled with all the patients' details, tumour location and radiological findings. Different techniques can be used: fine needle aspiration, core needle biopsy, incisional biopsy and excisional biopsy (Yaw 1999; Hogendoorn, Athanasou et al. 2010).

The location of the biopsy must be planned in detail getting information from the imaging finding since the region of the biopsy will also be included in the surgery plan during the resection of the tumour when the limb is saved in order to decrease the risk of local recurrences. Moreover any structure next to the tumour that is infiltrated with tumour cells must be taken in order to take as many information as possible with the obtained tissue (Yaw 1999; Hogendoorn, Athanasou et al. 2010; Ritter and Bielack 2010).

In most of the cases core needle biopsy is performed; the open biopsy is chosen in difficult cases depending on the tumour location and on the technical skills but it is the methodology that assures the collection of sufficient material. An excision biopsy is quite risky because the chances of contamination are higher above all when the tumour is malignant (Yaw 1999; Hogendoorn, Athanasou et al. 2010).

1.3.4 HISTOLOGY

After biopsy the pathologist will analyse the tissue and the final staging is defined in a multidisciplinary team made of pathologists, oncologists, radiologists and orthopaedists. Histologically OS appears characterised by spindle-like cells that produce immature bone, so called osteoid that is not exclusively produced by osteoblasts as one would expect because there is no proof that the malignant cells originate from osteoblasts that de-differentiate. Further OS tumours produce different amount of fibrous tissue and cartilage matrix that makes the diagnosis more difficult (Klein and Siegal 2006).

In general the subtypes of OS can be defined according to the different histological subtypes (table 1.4).

Types	Subtypes	Differential diagnosis
Conventional OS	Osteoblastic	Osteoblastoma
	Chondroblastic	Chondrosarcoma
	Fibroblastic	Fibrosarcoma
Teleangiectatic OS		Aneurysmal bone cyst
Small cell OS		Ewing's sarcoma/osteomyelitis
Low grade central OS		Fibrous dysplasia
Parosteal OS		Osteochondroma
Periosteal OS	High grade surface OS	Chondrosarcoma
		Periosteal chondrosarcoma
Secondary OS	Paget's disease	
	Post-irradiation	

Table 1.4: Classification of OS types and subtypes according to the WHO, on the basis of histological diagnosis, localisation and aggressiveness of the tumor (adapted from (Wirth and Winkelmann 2004)).

I will briefly summarise the most important histological features of the most common OS types.

The conventional OS presents with spindle or polyhedral cells with pleomorphic and hyperchromatic nuclei. The tumour cells produce bone or osteoid and are surrounded by extracellular matrix of osseous, cartilaginous or fibrous origin: usually all the three matrices are present in the histology but depending on which is prevalent the name of the histological subtype can be osteoblastic, chondroblastic or fibroblastic (Fig.1.5) (Klein and

Siegal 2006). The most common one is the osteoblastic (50%) while chondroblastic and fibroblastic are less recurrent (25% each) and all the three subtypes account for 80% of all OS (Fletcher 2002; Carrle and Bielack 2006).

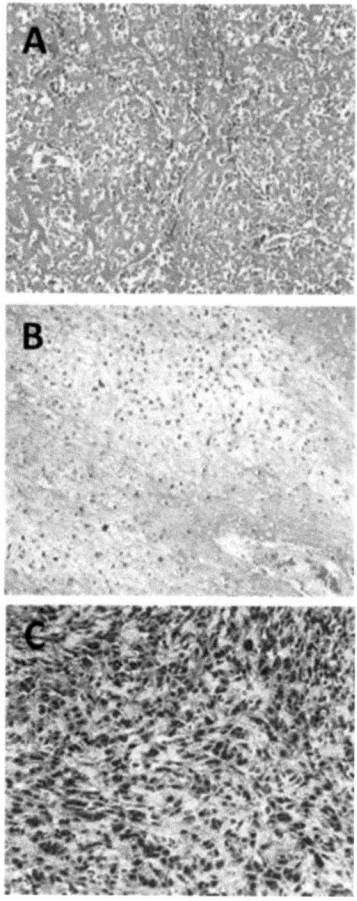

Fig.1.5: Histological pictures of osteoblastic (A), chondroblastic (B) and fibroblastic (C) OS tumour tissue section (Klein and Siegal 2006).

The telangiectatic type accounts for 4% of OS. It is often confused with the aneurysmal bone cyst since this OS type is characterised by blood-filled or empty spaces separated by a thin layer of atypical tumour cells. Radiologically the bone looks destroyed and the lesions are mainly radiolucent therefore it is rather difficult to visualise new bone formation in the x-rays. Histologically the osteoid is visible on the tumour sections but it is much more spread and forms a fine layer (Fig. 1.6) (Dorfman 1998; Fletcher 2002).

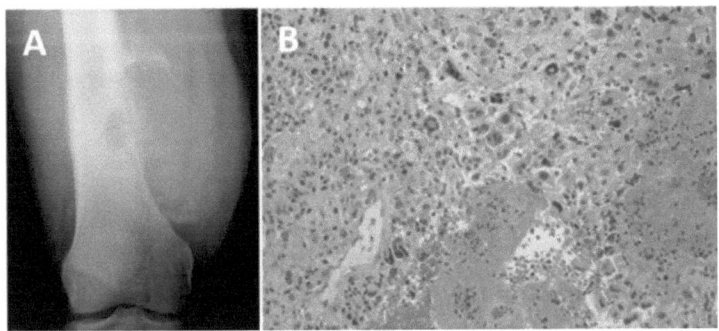

Fig.1.6: Example of a radiographical (A) and histological (B) appearance of telangiectatic OS (pedorthpath.com adapted from (Dorfman 1998; Fletcher 2002)

Small cell OS represents 1.5% of all OS; it affects usually female in the second decade of life. It is radiologically characterised by the presence of osteolytic lesions mixed to radiodense regions and mineralised tissue that extends in the soft tissue tumour or intramedullary (Fig.1.7). Histologically it displays presence of small cells surrounded by osteoid with round to oval nuclei and since the diagnosis is rather problematic because it can be mistaken with Ewing's Sarcoma, usually diagnosis is based on cytogenetic study for a traslocation t (11,22) that is usually not present in this OS type.

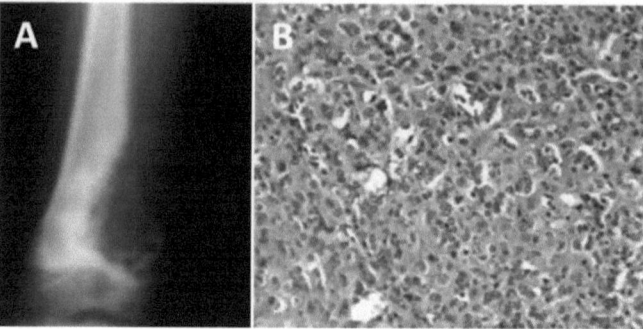

Fig.1.7: Example of a radiographical (A) and histological (B) appearance of small cell OS (Klein and Siegal 2006)

Low grade Central OS is another rare type (1-1.2%) that histologically appears similar to fibrous dysplasia and other benign lesions characterised by fibroblastic stroma surrounded by osteoid with spindle-like cells that invade the bone marrow and the bony trabeculae (Fig. 1.8). Imaging findings are not helpful for a final diagnosis but can highlight the more aggressive phenotype of this tumour compared to a dysplasia or other benign lesions (Fletcher 2002; Klein and Siegal 2006).

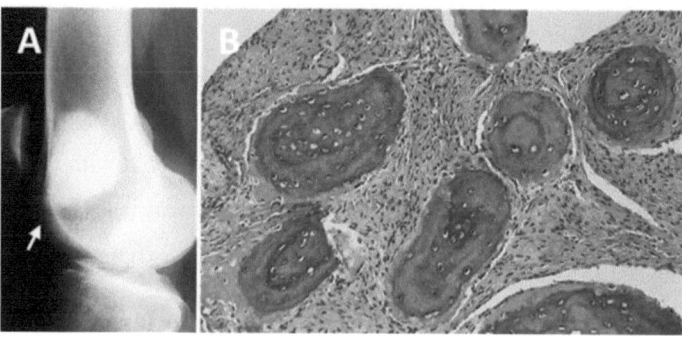

Fig.1.8: Representative x-ray (A) and histology (B) pictures of low grade central OS

Parosteal and periosteal OS are other uncommon types of surface OS that is more frequent in patients in the third or fourth decade of life. These OS types do not metastasise but show high recurrence rate.

The parosteal in 75-80% of the cases show radiopaque masses close to the distal posterior femur that are much denser in the centre and become less dense in the periphery (Klein and Siegal 2006; Yarmish, Klein et al. 2010). Finally the features of the periosteal OS are: presence of cartilaginous matrix and de-differentiated tumour component like in a moderately differentiated chondroblastic OS; moreover the periphery of the tumour shows absence of calcified bone and presence of spindle-like cells (Fletcher 2002; Klein and Siegal 2006).

1.3.5 STAGING

According to the WHO, the staging of tumours depends on the size of the tumour (T), on the metastatic spread (M) and on the number and distance of lymph nodes (L) affected by the tumour cells. Here below you see the list of the TNM criteria (table 1.5) and the staging (table 1.6) classification (Ritter and Bielack 2010).

Primary Tumour	Metastases	Lymph-nodes
TX not evaluable	MX not evaluable	NX not evaluable
T0 not detectable	M0 no distant metastases	N0 no lymph-nodes involvement
T1 size ≤ 8cm in the greatest dimension	M1a distant metastases in the lung	N1 regional lymph-nodes involvement
	M1b distant metastases in other organs	
T2 size > 8cm in the greatest dimension		
T3 discontinuous tumour		

Table 1.5: Definition of the criteria for tumour staging according to the tumour size, metastatic spread and lymph-nodes involvement (Adapted from (Ritter and Bielack 2010)

Stage IA	(low grade)	T1	N0	M0
Stage IB	(low grade)	T2	N0	M0
Stage IIA	(high grade)	T1	N0	M0
Stage IIB	(high grade)	T2	N0	M0
Stage III	(any grade)	T3	N0	M0
Stage IVA	(any grade)	Any T	N0	M1a
Stage IV B	(any grade)	Any T	N1	Any M
		Any T	Any N	M1b

Table 1.6: Staging of OS according to the TNM criteria described in table 5 (Adapted from (Ritter and Bielack 2010)). The difference in A and B depending on the size of the tumour in stage I and II and intracompartmental (A) or extracompartmental (B) location of the tumour in stage IV (Clark, Dass et al. 2008).

1.4 PROGNOSIS AND TREATMENT

Before the seventies, the prognosis was really bad: despite the local control was successful after primary tumour resection, OS patients were dying because of pulmonary metastases. After the seventies the survival increased to 68% in female and male patients with a localised disease and the death rates was decreasing 1.3% per year demonstrating that the combination of surgery and neoadjuvant chemotherapy was giving satisfactory results (Carrle and Bielack 2006; Savage and Mirabello 2011). Patients with metastatic disease reach a 5-year survival rate of only 20%; therefore the metastatic treatment still remains a challenge for researchers that need to focus on more specific therapies to fight against the more resistant and chemoresistant tumour cells (Lamoureux, Trichet et al. 2007).

Before going into the detail of the treatment protocol, I will summarise the most important clinical factors that are also affecting the prognosis of OS patients.

1.4.1 CLINICAL FACTORS

The most important clinical factors in OS are: tumour localisation, tumour stage, patient age, tumour size, pathological fracture, surgery, local recurrence, metastases, chemotherapy response and sera markers.

As mentioned in the paragraph concerning *"Incidence and Etiology"*, distal femur and proximal tibia are the most common sites of OS growth and in both sites the prognosis is quite good with a 5-year survival of 66% for the distal femur and 77.5% for the proximal tibia. In a study performed by Wittig, 23 patients with OS in the humerus showed a 10 years-overall survival of 65% which does not differ too much from the other two most common sites of OS development (proximal tibia and distal femur) (Wittig, Bickels et al. 2002). The good prognosis in these three common locations might be related to the high production of anti-angiogenic factors like PEDF in the growth plate and in the articular cartilage at the knee. The best prognosis is for patients with OS in the radius that show a 5-year overall survival of 81.3% while the worst is in pelvis between 27 and 47% and in the spine where the median survival is 1-2 years. Interestingly in both cases patients showed an increased survival when surgery was followed by radiotherapy (Bielack, Kempf-Bielack et al. 2002; Clark, Dass et al. 2008). Another prognostic factor is tumour staging: the stages IA and IB (table 1.6) show almost 100% of 5-year overall survival while already the stage II B, which is most commonly diagnosed, gives a much worse survival (47%) even when the patients are treated before surgery with chemotherapy (Foukas, Deshmukh et al. 2002). Patients with a stage III at diagnosis show lung metastases according to the Enneking classification system and a 5-year overall survival of 68%. According to the WHO classification and the American Joint Committee on Cancer patients stage III includes only patients with skip metastases (table 1.6) while patients with pulmonary metastases are classified in stage IV A and B (Clark, Dass et al. 2008).

Concerning the patient age, there are many studies highlighting that the prognostic factor is really bad in children or younger than 14 or older than 40 because of highest rate of metastatic incidence and a lower tolerance to chemotherapy in these patients (Bielack, Kempf-Bielack et al. 2002; Aksnes, Hall et al. 2006; Clark, Dass et al. 2008).

Primary tumour size as a prognostic factor has been also considered but so far only one study, performed by Bieling et al., gave some threshold and values to be considered. In this study on 128 patients they calculated the primary tumour volume with a specific ellipsoid formula and they found a value of 150 cm^3 to be the threshold above which the survival rate decreases from 92% to 58% (Bieling, Bielack et al. 1991).

Regarding the pathological fracture, a retrospective study compared the 5-year overall survival in a group of patients with pathological fracture with a group of people without and observed a higher 5-year overall survival (77%) in the patients that did not show any fracture compared to the group of patients with pathological fracture (55%) (Scully, Ghert et al. 2002).

This finding arose the question whether the different surgical options, amputation and limb salvage surgery, might affect OS patients survival. Contradictory are the results in this concern: Bacci et al. noticed no statistical difference between the two surgeries while Schrager et al. recently published a paper on the survival rates of OS patients comparing the limb salvage surgery with the amputation and interestingly patients that underwent a limb salvage surgery showed a 5-year overall survival of 72.7% compared to only 60.1% in case of patients that underwent amputation (Bacci, Ferrari et al. 2003; Schrager, Patzer et al. 2011).

What is important to mention is that amputation is usually performed in patients with advanced diseases, therefore the inverse prognosis does not depend on the amputation per se. Neverthless Abudu et al. observed that the limb salvage surgery has a slighltly increased risk of local recurrence compared to the amputation (Abudu, Sferopoulos et al. 1996).

Local recurrence is one of the key prognostic factor because it is usually associated with a more aggressive phenotype of the tumour cells and to a high probability to develop pulmonary metastases. Indeed it was visible that patients with local recurrence and metastases have a much lower survival (4%) compared to patients with only metastases (22%) (Bacci, Donati et al. 1998).

From this evalutation, we can conclude that the combination between local recurrence and metastases is negative.

Concerning metastases as a prognostic factor alone, usually bone metastases give a much worse survival than lung metastases. In the case of lung metastases usually the patients that present with metastatic foci after chemotherapy had a poorer survival since the one showing metastases already at the diagnosis underwent a metastasectomy via a radical thoracotomy procedure which is much more sensitive than metastases detection through CT. Finally skip metastases, which occur as local metastases along the length of the bone, are quite rare in OS patients: only 1.8-25% of OS patient show skip metastases with a poor overall survival of 27.2 months from diagnosis (Clark, Dass et al. 2008).

Another important prognostic factor is the response to chemotherapy: this effect is evaluated from the percentage of necrosis that the tumour section overall shows. The threshold for good and bad response is set at 90% and usually the good responders show more than 90% of necrosis and they usually also have a better prognosis (Clark, Dass et al. 2008). Interestingly the treatment response of patients depends strongly also on the histological type of OS that they display: patients with teleangiectatic type show the best prognosis (86.7%), while patients with small cell tumours show the worst response (25%). Good response is also achieved by patients with osteoblastic (63.9%), fibroblastic (61.7%) and chondroblastic (50.6%) (Bacci, Longhi et al. 2006).

The last prognostic factor that will be described is alkaline phosphatase (ALP) which is correlated, when high before treatment, to a worse 5-years overall survival (54%) compared to patients with low ALP level in the serum (67%) (Clark, Dass et al. 2008).

1.4.2 TREATMENT

Originally the standard protocol for OS treament consisted of the use of pre-chemotherapy which lasts around 10 weeks followed by surgical resection of the primary tumour and 20 weeks of neo-adjuvant chemotherapy (Chou and Gorlick 2006). The most effective and so far used drugs in OS are high-dose methotrexate (HD-MTX), adriamycin and cisplatin.

In 2001 a new project started with the name of EURAMOS1 which was the result of a collaboration between the European and the North American Society. This joint effort is needed in such a rare tumour since this makes possible to reach statistically significant results involving a higher number of patients: from 1st April 2005 till 30th June 2011, 2260 patients were registered and 1332 were randomised making this the largest clinical trial ever existing for OS. Their main interest is to increase the overall survival improving the therapeutic strategies on poor responders. The main aims of this project are two: evaluate whether ifosfamid and etoposide could help patients with a poor prognosis 10 weeks after the pre-chemotherapy and whether the pegylated formulation of the interferon α (INF α) increases the overall survival of patients with good response after the pre-chemotherapy (Chou and Gorlick 2006).

The scheme of the treatment defined by EURAMOS1 consists of a pre-chemotherapy phase where HD-MTX, adriamycin and cisplatin are used, surgery and post-chemotherapy where depending on the histological response, patients will receive randomly either an addition of INF α or not in case of good response, and ifosfamide and etoposide or nothing in case of poor response (Fig. 1.9). Pre-surgery chemotherapy treatment is strongly recommended because the chemotherapy will induce a clear demarcation of the tumour borders through the formation of an avascular capsule around the tumour cells. Furthermore the pre-treatment allows to get information about the histological response before the surgery so that the patient can be directed towards different therapeutical strategies depending on the percentage of necrosis that the tumour shows (Carrle and Bielack 2006).

The three drugs, included in the pre-treatment, were found to be effective around 1980s and since then, no additional drug has been found to be more efficacious. Concerns are present for the side effects of the adriamycin and cisplatin on the heart, ears and kidneys and for the real need of HD-MTX in the treatment regimen (Bielack, Kempf-Bielack et al. 2002). A German and Brasilian study on the HD-MTX showed that the overall survival did not differ much in the presence or absence of HD-MTX. Despite these results, it is still maintained in most of the OS regimens for the relative lack of myelotoxicity and late effects: side effects from HD-MTX are linked usually to nephrotoxicity and can be reduced using the glucarpidase that cleaves the MTX and leucovorine which is an analogue of folic acid and avoid cell death due to a lack of folic acid caused by the MTX (Carrle and Bielack 2006). After the pre-chemotherapy that can last differently depending on the tumours, the patients undergo the surgery. Till the 1970s the amputation was the only choice for OS patients but now limb salvage surgery is performed in 85% of the cases and it consists of primary tumour resection including a wide area of margins and reconstrunction of the missing bone structures. There is a slight higher incidence of tumour recurrence in the limb salvage surgery which is linked to the resection of the margins as discussed in the sub paragraph about *"Clinical Factors"* but this concerns only the patients with close margins of excision where amputation can be preferred.

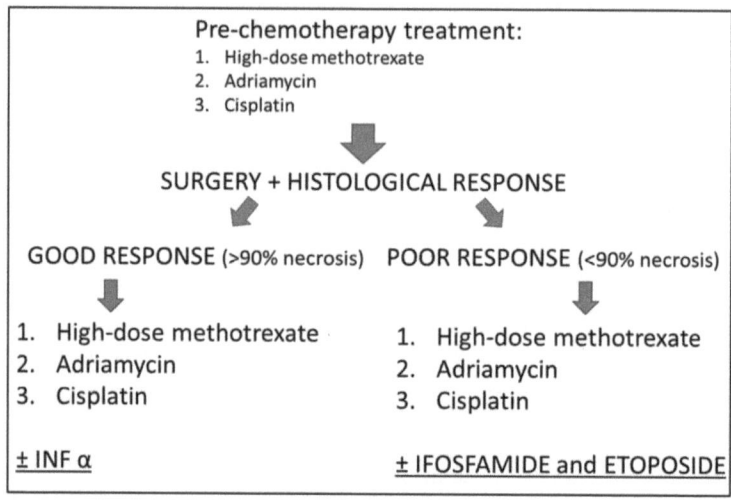

Fig.1.9: Scheme of the treatment protocol according to the EURAMOS clinical trial; (IFN α= interferon α) (adapted from (Bielack, Carrle et al. 2008)).

Moreover the limb salvage surgery is more demanding for the following reconstruction of the bone above all in young adolescence where the excision of the growth plate from the surgery causes a decreased growth (Abed and Grimer 2010). After the surgery the maintenance therapy includes the use of INF α in case of histological good response evaluated on the biopsy, or of ifosfamide or etoposide in case of poor response.

IFN α is a cytokine involved in anti-viral, anti-tumour, apoptotic and cellular immune response activities. The anti-tumour effect was discovered by Gresser on mice with a non- viral dependent tumour. In OS the INF α effect was demonstrated *in vitro* on human OS cells and *in vivo* on four different xenograft mouse models of OS (Strander and Einhorn 1977;

Whelan, Patterson et al. 2010). Clinically INF α was used in one clinical trial performed in Scandinavia where they tested the efficacy of INF α on 89 patients in two series and they found that patients treated with higher dose showed a higher survival compared to the ones treated with a lower dose proposing a dose-response effect (Muller, Smeland et al. 2005). In the EURAMOS1 study the INF α is given in a pegylated form which strongly reduces the side effects combined to it (Whelan, Patterson et al. 2010).

Regarding ifosfamide, a cooperative German-Swiss-Austrian study showed that the inclusion of ifosfamide in the treatment regimen for OS patients increased the 10-year overall survival to 78% in case of a localised disease (Fuchs, Bielack et al. 1998). The pitfall of this study was the low number of patients, therefore the real benefit of ifosfamide will be assessed in the EURAMOS1 study. Another study, finally, highlighted the importance of the combination of ifosfamide and etoposide in the management of patients with primary and metastatic OS (Goorin, Harris et al. 2002).

Beside EURAMOS 1, another collaboration between the Italian and Scandinavian Sarcoma Groups was founded with the name of EURO-B.O.S.S. that includes patients over 40 with bone sarcomas. In this joint effort the main focus of the clinical trial is to analyse the efficacy of the treatment based on the use of adriamycin, cisplatin and ifosfamide. HD-MTX is included in the regimen only in cases of extreme poor response since in elder patients the toxicity of this drug is much higher. Furthermore the used dosage in these cases is lower $8g/m^2$ compared to the $12g/m^2$ used in young patients.

Radiotherapy in OS is rarely used and only in cases where the resection is difficult to be performed (Bielack, Carrle et al. 2008).

Finally in case of metastatic disease, the overall survival is much lower as described in the paragraph *"Clinical factors"* and the treatment remains the same and in addition surgical resection of the metastatic foci has to be performed. Usually around 30% of the patients with resected metastases can survive longer than 5 years and even in case of multiple operations, patients can have higher probability to survive as long as metastases can be resected (Bielack, Kempf-Bielack et al. 2009; Ritter and Bielack 2010). The benefit from the use of chemotherapy in case of recurrent OS is not yet determined (Ritter and Bielack 2010).

Monitoring of the treatment in OS is performed through metabolic imaging because anatomical differences of the tumour size can only be visualised at a later time point. Indeed CT and Magnetic Resonance Imaging (MRI) do not give information about the necrotic and viable parts of the tumour. On the other side it is well accepted that Positron Emission Tomography (PET) using ^{18}F-FDG as a tracer is the most sensitive tool for evaluation of treatment response but the importance of PET imaging in OS will be analysed in more detail in the next chapter (Messa, Landoni et al. 2000). After the treatment, local or metastatic relapse is followed up each 6 weeks to 3 months in the first 2 years after OS diagnosis; each 2-4 months for the next 2 years and every 6 months until 5 years and thereafter yearly. The follow-up strategy includes a physical examination, a chest X-Ray to check for lung metastases and if the X-Ray finding is unclear, chest CT is performed.

1.4.3 CHEMORESISTANCE IN OS

Chemoresistance is a major problem for the treatment of tumours because it is the main cause of poor response in OS patients and consequently new therapeutic approaches have to be tried and investigated which could effectively attack the resistant tumour cells. Chemoresistance can be intrinsic when it is from the beginning of the treatment and it might depend on pathogenic mechanisms or it can be acquired during therapy and in this case it is usually caused by accumulation of genetic mutations from some cells that, consequently, gain the ability to resist to the treatment and survive. Cisplatin, Adriamycin and Methotrexate induce the activation of specific mechanism that causes the efflux of the drug from the tumour cell (Chou and Gorlick 2006).

Resistance to Cisplatin, for example, is usually acquired through the overexpression of multidrug resistance-associated protein 2 that is responsible of a decreased concentration of cisplatin in the cell (Borst, Evers et al. 2000; Itoh, Tamai et al. 2002). Further it was shown that cells after Cisplatin crosslinks with DNA proceed in the replication without undergoing cell death or repair mechanism. This is usually dependent on p53 acquired mutations (Manic, Gatti et al. 2003).

In the case of Adriamycin, the resistance is dependent on the overexpression of the p-glycoprotein that might also occur after p53 gets mutated though the results confirming the correlation between the two events are conflicting (Serra, Scotlandi et al. 2003; Tsang, Chau et al. 2003). A possibility to overcome this resistance would be the use of increased doses of Adriamycin with the time and *in vitro* it was shown that type I interferon decreases the expression of p-glycoprotein in resistant OS cells (Manara, Serra et al. 2004).

Finally Methotrexate resistance is well studied and occurs in the different steps of the metabolic pathway of this drug which is an analogue of folic acid: impaired transport of methotrexate caused by the lower expression of the folate carrier; mutations in the dihydrofolate reductase gene which reduces the affinity of the methotrexate for the enzyme which is the main target of the drug; increased drug expulsion through the ABC family of transporters (Chou and Gorlick 2006).

CHAPTER 2

INTRODUCTION: POSITRON EMISSION TOMOGRAPHY AND PHOTODYNAMIC THERAPY

Chapter 2 Introduction: Positron Emission Tomography and Photodynamic Therapy

2.1 POSITRON EMISSION TOMOGRAPHY

Positron Emission Tomography (PET) is a non invasive imaging tool that allows the evaluation of metabolism within human body. It was established in the late 1970s but only in the 1990s it started to be broadly used (Phelps, Hoffman et al. 1975; Weber 2006).

It differs from CT and MRI because it does not give information about the anatomy and from single-photo emission tomography (SPECT) because with PET data can be quantified. Furthermore the resolution and the sensitivity of PET scanner are much higher compared to the SPECT and compared to biopsies PET imaging is not invasive and gives much more information over time about the changes in the tumour development and volume. The main pitfall is the spatial resolution which is around 4-10 mm: despite the improvements over the last years, it is rather difficult to detect small lesions and heterogeneity of the tumour in larger lesions (Kelloff, Hoffman et al. 2005; Weber 2006; Fass 2008). Therefore more and more the combination of PET/CT is used in order to overcome the limitations of the single techniques: in this way the information obtained are complete and include functional (PET) and anatomic (CT) data. The principle of PET imaging is based on the use of positron emitting radioisotopes, such as oxygen-15 (^{15}O), carbon-11 (^{11}C), nitrogen-13 (^{13}N) fluorine-18 (^{18}F) and zirconium-89 (^{89}Zr). These radionuclides are the most common because they can easily be linked to molecules that resemble biological substances like ^{18}F-Fluorodeoxyglucose; drugs like ^{89}Zr-DN30 antibody directed against cMet; exogenous molecule like ^{18}F-Fluoromisonidazole that targets hypoxic areas (Hong, Zhang et al. 2012).

^{15}O, ^{11}C and ^{13}N do not affect the function and activity of the molecule they will react with because they are replacing the same atom in the molecule. In the case of ^{18}F, the radioisotope substitutes the hydrogen since there are no radioisotope for hydrogen therefore it is important that the substitution of hydrogen with ^{18}F does not have any impact on the biological function of the molecule. Another criteria for the selection of the proper radionuclide is the half-life which is the time needed to decay and reduce to half the initial dose injected in the patient. On one side when the half-life is short like in the case of ^{15}O and ^{13}N, then the center should have a cyclotron inside because the study has to be performed in short time (table 2.1). On the other side when the half-life is long such as in the case of ^{89}Zr then it is also not so convenient because this increases the patient waiting time for the radioactive decay. This is also the reason why most of the PET imaging centers use ^{18}F as the most common radioisotope for molecules labelling (Phelps 2000; Schlyer 2004).

Radioisotope	Half-life
Carbon-11	20.38 min
Nitrogen-13	9.97 min
Oxygen-15	2.04 min
Fluorine-18	109.8 min
Zirconium-89	78.4 hrs

Table 2.1: Half-lifes of the most commonly used radionuclides

The radioisotopes are characterised by the presence of more protons and less neutrons compared to a stable nucleus and therefore they release positrons which are positively charged and annihilated with electrons giving rise to two photons with the same energy (511keV) but opposite directions, 180 degrees from each other (Gopal 2003; Kelloff, Hoffman et al. 2005; Ametamey, Honer et al. 2008).

The PET instrument consists of a ring of detectors that are collecting only the coincident photons that reach the opposite detectors at the same time. After the chemical binding of the radioisotopes with our targets that highlight the metabolic processes we are interested in, the tracer is intravenously injected and after a waiting time for the tracer distribution the patient is set in the ring and the information about the coincident events are sent to the computer for the image reconstruction. Here below you can see a schematic view of micro-PET on a mouse using ^{18}F-Fluoride as a radioisotope (Fig. 2.1).

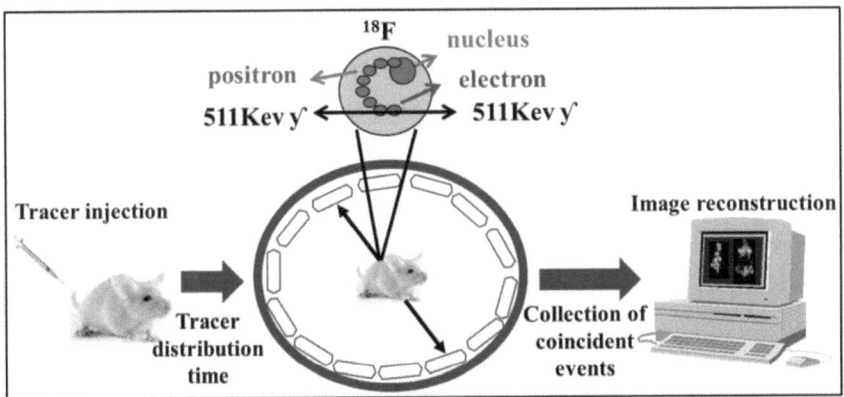

Fig.2.1: Scheme of micro-PET in mice using ^{18}F as a positron emitting radioisotope (adapted from www2.fr-jurglich.de/zel/zel_bildgebung_beschreibung)

2.1.1 PET IN THE CLINICS

During the last years the development of PET imaging in the pre-clinical or clinical field strongly increased. Some of the PET applications will be discussed below.

^{18}F-Fluorodeoxyglucose (^{18}F-FDG) is the most used PET tracer in the clinic and it is a measure of glucose metabolism in the human body. As a glucose analogue, it is transported into cells through glucose transporters (GLUTs). After being phosphorylated at position C-6, the 2-deoxyglucose-6-phosphate is trapped in the cell since it can proceed neither in the glycolytic pathway nor in the glycogen synthesis. Background uptake is visible in organs that physiologically use glucose as a source of energy such as muscle, heart and

Chapter 2   Introduction: Positron Emission Tomography and Photodynamic Therapy

brain but also it accumulates in the bladder because of excretion. Therefore it is not recommended to use ^{18}F-FDG for the evaluation of alterations in these organs (Schlyer 2004).

Indeed in the case of brain studies, for example, neutral aminoacids are usually used to image the brain because of the lower physiological uptake. Indeed the most produced radionuclides for the brain imaging are ^{11}C-methionine, ^{11}C-tyrosine and ^{18}F-ethyltyrosine or ^{18}F-methyltyrosine. The main application for the brain imaging is the tumour delineation in preparation for radiotherapy (Grosu, Weber et al. 2005; Grosu, Weber et al. 2005; Weber 2006).

Nevertheless ^{18}F-FDG use is strongly suggested to define the damage induced by the ischemia and together with a measurement of the perfusion it is used to assess the efficacy of therapy. In this context also new tracers have been developed such as nitrogen-13 ammonia which is specific for the evaluation of the blood flow (Schlyer 2004).

Further PET imaging helps in studying the ligand-receptor interactions. Indeed nowadays it became common to label the drug with a specific radioisotope and get information about the specificity and the distribution of the new drug using PET (Knuuti and Bengel 2008). The first study was performed by Shani and Wolf who labelled a new drug, the Fluorouracil, with 18-Fluorine. With PET imaging they demonstrated that ^{18}F-Fluorouracil is helpful in distinguishing the tumours which are sensitive to the treatment from the resistant ones though it was rather complicated to quantify the tumour accumulation because of quick excretion and/or accumulation in other organs (Shani and Wolf 1977). Furthermore it is always difficult to label the drug without affecting its function. Despite these challenges, PET imaging with ^{18}F-Fluorouracil is useful to quantify the tumour accumulation of this tracer when treatment with eniluracil, that inhibits the processing of Fluorouracil, was performed (Saleem, Yap et al. 2000). PET imaging gives also information about the accumulation of the drug in the tumour cell and the expression of multidrug resistance-associated protein

1 which are connected with treatment resistance because of drug expulsion from the cell (Weber 2006).

PET is also used to evaluate the effect of a therapy. One example is the imaging with reporter genes that usually are exogenous to the body but not toxic. Basically the therapeutic gene is either itself a reporter or linked through an internal sequence, called IRES, to a reporter gene that can be a transporter, an enzyme, a receptor and so on. There are not many studies on the use of such reporters for PET imaging: one study was successfully performed in patients with hepatocellular carcinoma. The aim of this project was to evaluate via PET imaging the efficacy of the use of a herpes simplex 1 thymidine kinase (HSV1-tk) overexpressing adenovirus to monitor and treat the hepatocellular carcinoma because the enzyme HSV1-tk, besides being a reporter gene, also kills the tumour cells via activation of a non-active pro-drug like penciclovir in an active and toxic compound. In this study they injected intratumourally the adenovirus and used 9-(4-^{18}F-Fluoro-3[hydroxyl methyl] butyl) guanine (^{18}F-FHBG), a penciclovir analogue, to image the area of the tumour since the PET tracer will accumulate only in the tumour cells where HSV1-tk is expressed (Alauddin, Shahinian et al. 2001; Penuelas, Mazzolini et al. 2005).

Besides reporter genes also receptors are successfully imaged through the radiolabelling of ligands like 16α-Fluoroestradiol-17β (FES) which is used for the visualisation of the estrogen receptor in estrogen-positive advanced breast tumour and the 16β-^{18}F-Fluoro-5-dihydrotestosterone (FDHT) which is specifically taken up by metastatic foci of androgen-dependent prostate cancer (Mintun, Welch et al. 1988; Larson, Morris et al. 2004; Dehdashti, Picus et al. 2005).

More detailed information about the use of PET imaging in oncology will be given in the paragraph below.

2.1.2 PET IN ONCOLOGY

The clinical application of PET in oncology developed only recently due to the improvement in sensitivity and resolution of the new scanners and due to the reimbursement approved by the Center for Medicare and Medicaid Services for the use of PET in staging and restaging of small-cell lung, esophageal, colorectal, breast, head and neck cancers and for the treatment response monitoring in breast cancer (Kelloff, Hoffman et al. 2005). The evaluation of a PET image consists usually of two parts: the first one is visual and subjective; the second one is more objective and based on the calculation of the standardised uptake value (SUV). In the case of treatment response the analysis consists of a comparison of the SUV before and after the treatment.

SUV is a semi-quantitative index to measure the accumulation of the tracer in a region of interest (ROI) measured in MBq/g normalised to the injected dose (ID) in MBq and the body weight (BW) of the patient in gram. The factors that strongly influence the SUV calculations are: plasma glucose level, definition of the border for the ROI is quite subjective; attenuation to correct the PET signal from the collection of non-coincident events is required; reconstruction methods can affect the noise and the resolution; time of acquisition after the tracer injection has to be standardised and the competition for the transport of the tracer and its biological analogue that uses the same transporter to enter the cell (Thie 2004; Peterson 2007).

It is, therefore, necessary to standardise all the different steps for the SUV calculation to be able to compare different values, for example in the case of the evaluation of treatment response, the uptake before and after treatment.

^{18}F-FDG has been the most commonly used PET tracer for 25 years. Meanwhile many other tracers visualising particular metabolic processes have been synthesised (Table 2.2). Below, some of these tracers will be shortly described in the context of potential applications in the oncology field.

PET tracer	Targeted metabolic process	Application
^{18}F-Fluorodeoxyglucose	Glucose metabolism	Diagnosis, staging, restaging and treatment monitoring
^{11}C-Thymidine ^{18}F-Fluorothymidine	Proliferation	Diagnosis, staging, restaging and treatment monitoring
^{11}C-Methyonine	Protein synthesis	Diagnosis, staging, restaging and treatment monitoring
^{11}C-Choline ^{18}F-Fluorocholine	Cell membrane turnover	Staging, restaging and treatment monitoring
^{11}C-Tyrosine ^{18}F-Fluorotyrosine ^{18}F-Fluoroethyltyrosine	Protein synthesis	Staging, restaging and treatment monitoring
^{18}F-Fluorodihydroxyohenylalanine	Dopamine and protein synthesis	Staging, restaging and treatment monitoring
^{18}F-Fluoromisonidazole	Hypoxia	Evaluation of hypoxia in the tumour regions
^{18}F-Fluoro-17-β-estradiol	Estrogen activating receptors	Evaluation of positive estrogen receptors and monitoring of the treatment in estrogen-receptor-positive breast cancer
^{18}F-Annexin V	Apoptosis	Evaluation of hypoxia and treatment monitoring
^{18}F-Fluorouracil	Tumour tissue accumulation	Treatment monitoring
^{11}C-Acetate	Lypid synthesis	Staging, restaging and treatment monitoring

Table 2.2: List of the most common tracers for PET imaging with the targeted process and the main applications (Juweid and Cheson 2006)

2.1.1.1 PET TRACERS: *^{18}F-Fluorodeoxyglucose (^{18}F-FDG)*

^{18}F-FDG enters the cell through GLUT-1 and accumulates in the cell after being phosphorylated by hexokinase, as mentioned previously. Tumours show an increased uptake because of an increased rate of glycolysis that occurs also in absence of oxygen (Warburg effect) which is in addition linked to an increased expression of hexokinase and GLUTs. According to Gambhir et al the ^{18}F-FDG sensitivity and specificity in cancer reach the 84% and the 88% (Gambhir, Czernin et al. 2001). ^{18}F-FDG imaging is also used for staging in melanoma, lymphoma, esophageal and colorectal cancers (Kelloff, Hoffman et al. 2005). An exception to this is the prostate where the ^{18}F-FDG uptake is physiologically high therefore it is not recommended to use PET imaging for staging and restaging of this tumour (Kelloff, Hoffman et al. 2005; Juweid and Cheson 2006).

Concerning treatment monitoring, it was shown in patients with breast cancer that an accurate evaluation of treatment response could be already evaluated one cycle after chemotherapy and this is extremely important in order to change and define a better therapeutic approach in cases of poor response (Juweid and Cheson 2006).

Furthermore Weber et al. state that in the case of gastroesophageal cancer a reduction of 35% in the SUV of ^{18}F-FDG two cycles after chemotherapy has a prognostic value. Patients with a metabolic response display a median time to progression or recurrence was 16 months compared to the 9 months in the case of poor metabolic responders (p=0.01): differences in the overall survival were also significantly different (p=0.05) (Weber, Ott et al. 2001).

In sarcomas a study conducted with 46 patients shows that ^{18}F-FDG-PET was more sensitive than CT, MRI, bone scintigraphy and ultrasound in detecting primary tumour, lymph node involvement and in bone invasion but CT was more sensitive in detecting lung metastases (100% compared to the 25%) and these findings were confirmed also by other retrospective studies (Franzius, Sciuk et al. 2000; Franzius, Daldrup-Link et al. 2001; Iagaru, Chawla et al. 2006).

Another study performed on 21 patients with soft tissue sarcoma demonstrated that ^{18}F-FDG-PET responders showed a higher progression-free survival (92%)

compared to the non-responders (12%) but since the number of patients in this study is low, no significant conclusion can be extrapolated (Stroobants, Goeminne et al. 2003).

In another prospective study on 46 patients with localised high-grade soft tissue sarcoma a reduction in SUV that was higher than 40% obtained before and 2 cycles after chemotherapy correlated with an improved recurrence free and overall survival (Schuetze, Rubin et al. 2005).

Though ^{18}F-FDG is the most commonly used tracer, there are some drawbacks that induced the chemists to synthesise new tracers: ^{18}F-FDG is highly accumulating in inflammation due to the high uptake of the white blood cells; high background is present in muscle and grey matter due to the physiological need of glucose. Furthermore ^{18}F-FDG is not helpful in distinguishing the malignant from the benign pathologies and the uptake strongly depends on the blood flow and on the concentration of insulin (Lindholm, Minn et al. 1993; Culverwell, Scarsbrook et al. 2011). In diabetic people the use of PET imaging can give unreliable results due to the hyperglycaemia that can reduce the uptake of ^{18}F-FDG in some cancers like pancreatic and lung (Diederichs, Staib et al. 1998; Torizuka, Zasadny et al. 1999).

2.1.1.2 PET TRACERS: ^{18}F-Fluoromisonidazole (^{18}F-FMISO)

^{18}F-FMISO is the most used tracer for hypoxia. It passively diffuses through the membrane in the cell and the -NO$_2$ group, which has high affinity for electrons, reacts with an electron to form a radical anion and afterwards if oxygen is present, the electron will be transferred to the oxygen otherwise a second electron can react with the nitroimidazole group and form a product that will react further with peptides and RNA and therefore gets trapped and accumulates in the cell. In necrotic areas the ^{18}F-FMISO is not trapped because the electron transport chain is not working and therefore the electrons cannot be transferred (Imam 2010).

Hypoxia is a phenomenon occurring in many pathological conditions, including stroke, tissue ischemia, inflammation and solid tumours due to a non-effective transfer of oxygen from the vasculature to these areas. Despite the reduced level of oxygen is cause of tumour cell death due to lack of nutrients transfer, some tumour cells can acquire mutations which enables them to survive and increase their malignant features. As a consequence, hypoxia areas in the tumour can show more resistance to treatment and therefore it is needed to investigate in advance the hypoxia status of the tumour in order to apply, later, a more specific treatment schedule (Wilson and Hay 2011).

As a hypoxia marker, ^{18}F-FMISO PET imaging is preferred to the *Gold Standard* method which uses directly the oxygen electrode but that it is quite demanding and needs high expertise.

It is showed to be successfully used for evaluation of the prognosis after radio- and chemo-therapy. Rajendran et al. reported that ^{18}F-FMISO can be used as an independent prognostic factor after using it in 73 patients with head and neck squamous cell carcinoma (Rajendran, Schwartz et al. 2006). Same findings were shown by other groups that worked on head and neck cancer and non-small cell lung cancer (Hicks, Rischin et al. 2005; Rischin, Hicks et al. 2006).

In sarcomas ^{18}F-FMISO was used only in animal studies and mainly in dog that developed spontaneously different kind of sarcomas. They could detect tumour hypoxia in the center of the tumour where the perfusion was low (Bruehlmeier, Kaser-Hotz et al. 2005).

2.1.1.3 PET TRACERS: ^{18}F-Fluoride

Among all the tracers, ^{18}F-Fluoride is the most specific one for bone imaging. It diffuses through the bone vasculature into the matrix and there it is trapped since it replaces the hydrogen in the hydroxyapatite. Its uptake strongly depends on the blood flow and on the osteoblastic activity. It can highlight benign and malignant bone lesions but there is no clear SUV cut off between malignant and benign conditions. The absence of tumour specificity is the main disadvantage and therefore PET imaging cannot give a final diagnosis. Despite the recent developments of ^{18}F-Fluoride imaging, it is not yet used routinely in the clinic but it may replace the bone scintigraphy in few years (Even-Sapir, Mishani et al. 2007).

Most of the studies with ^{18}F-Fluoride were performed in patients with bone metastases from breast and prostate cancer. Bone metastases in these patients are osteoblastic, osteolytic or a mixture of both and it was found that osteoblastic lesions take up more ^{18}F-Fluoride than osteolytic ones. Bone scintigraphy is also applied for diagnosis of bone metastases but PET imaging is more sensitive than bone scintigraphy therefore its use is strongly suggested (Langsteger, Heinisch et al. 2006).

2.1.1.4 PET TRACERS: ^{18}F-Fluorodeoxythymidine (^{18}F-FLT)

^{18}F-Fluorodeoxythymidine (^{18}F-FLT) is an analogue of thymidine and it is used as a marker of DNA replication since this nucleoside is the only one that is not part of RNA. The information about the tumour proliferation gives a hint for the evaluation of the tumour growth. Cellular proliferation is usually measured from biopsy, invasive method, that might be difficult to perform and the tumour biopsy usually does not represent the proliferation rate in the whole tumour. Therefore PET imaging with ^{18}F-FLT can address this question being less invasive and its uptake is visualised in the whole area of interest.

After being intravenously injected, ^{18}F-FLT is transported into the cells via Na$^+$-dependent nucleoside transporter in normal tissues and in part also by passive diffusion mainly in the tumour cells (Belt, Marina et al. 1993; Reske and Deisenhofer 2006). After entering the cells it is phosphorylated by thymidine kinase 2, which is expressed only in proliferating cells, and as FLT-triphosphate it can be incorporated in the DNA and accumulates in the cytosol. Since ^{18}F-FLT phosphorylation depends mainly on the expression of the thymidine kinase, the ^{18}F-FLT accumulation relates in most of the cancers to the proliferation rate in the tumour area (Barwick, Bencherif et al. 2009).

Some preclinical studies report that ^{18}F-FLT can be used to measure the treatment response at earlier stages and they confirmed the result from the PET imaging with the immunostaining for the proliferating cell nuclear antigen (PCNA) (Sugiyama, Sakahara et al. 2004; Yang, Ryu et al. 2006; Molthoff, Klabbers et al. 2007).

In the clinic, ^{18}F-FLT is used mainly in brain tumours because compared to ^{18}F-FDG it displays a low uptake in physiological conditions and also in breast, oesophageal, lung, lymphoma and sarcomas but in these other tumours the ^{18}F-FLT uptake is lower than ^{18}F-FDG thus limiting the use of ^{18}F-FLT for tumour staging. Furthermore since the physiological uptake of ^{18}F-FLT in the liver and in the bone marrow is moderate to high, it is not used to detect liver and bone marrow's metastases (Barwick, Bencherif et al. 2009).

In relation to the treatment response, few studies were performed among which one compared the ^{18}F-FLT uptake before and one week after treatment in 13 patients with breast cancer and they found a significant correlation between the clinical response and the PET results (Kenny, Coombes et al. 2007). Same findings were obtained in a study performed on patients with advanced adenocarcinoma of the lung and with advanced rectal cancer (Barwick, Bencherif et al. 2009).

2.1.1.5 PET TRACERS: ^{18}F-Fluoroethyltyrosine (^{18}F-FET)

^{18}F-Fluoroethyltyrosine (^{18}F-FET) is a marker for protein synthesis and it enters the cell through a Na$^+$-indipendent amminoacid transporter, which is highly expressed in tumour cells, and afterwards it does not enter the protein synthesis (Spaeth, Wyss et al. 2004). Very few studies are performed on this tracer and they focus on the use of this tracer in brain tumours since the uptake in the brain and in benign conditions are low and it gets quite high in tumours condition (Wester, Herz et al. 1999; Caroli, Nanni et al. 2010).

In summary this tracer has high potential for the differential diagnosis of brain malignant tumours (Langen, Hamacher et al. 2006).

2.1.1.6 PET TRACERS: ^{18}F-Fluorocholine (^{18}F-FCH)

^{18}F-Fluorocholine (^{18}F-FCH) is an analogue of choline, essential component for the synthesis of phospholipids in the cell membrane that accumulates in the tumour cells since it was shown by Swinnen et al. that tumour cells have an increased demand of fatty acid and phospholipids (Swinnen, Brusselmans et al. 2006). Similar to choline, ^{18}F-FCH is transported into the cell via specific transporters, is phosphorylated by the choline kinase but afterwards it accumulates in the cells to a low extent and becomes the substrate of the cytidylyltransferase and proceeds in the phospholipids synthesis (Kwee, DeGrado et al. 2007)

The first clinical use of ^{18}F-FCH was in prostate cancer in alternative to ^{18}F-FDG that shows a high background in this organ. Contradictory are the findings about the possibility to use ^{18}F-FCH for differential diagnosis between benign and malignant lesions.

Schmid et al. published a study on 9 patients where the use of ^{18}F-FCH PET/CT was not sufficient for the differential diagnosis of prostate hyperplasia or prostate cancer. In contrast to this finding, Kwee et al. found that in 17 patients with prostate cancer, malignant sextants showed higher SUV compared to the biopsy negative sextants (Mertens, Slaets et al. 2010).

Furthermore ^{18}F-FCH showed low sensitivity (10%) in detecting lymphatic nodes in 10 patients with prostate cancer and in 20% PET imaging showed false-positive results due to accumulation of ^{18}F-FCH in inflammatory lesions (Hacker, Jeschke et al. 2006).

Concerning distant metastases, Pelosi revealed from a study on 56 patients with prostate cancer, that the sensitivity of PET/CT was dependent on the level of PSA and highest (82%) when the PSA level was > 5ng/ml (Pelosi, Arena et al. 2008).

In brain tumours, ^{18}F-FCH could discriminate between high-grade gliomas for the different pattern of tumour uptake which was peritumoural, brain metastases, benign lesions and also between recurrences and radiotherapy-induced effect (Kwee, Ko et al. 2007).

In hepatocellular carcinoma, ^{18}F-FCH gave promising results in a small study on 14 patients with diagnosed carcinoma or recurrent disease where ^{18}F-FCH uptake could discriminate between single lesion, found in 9 patients, or multifocal in the other three patients. No study was performed on other liver cancers (Talbot, Gutman et al. 2006).

Finally so far the use of ^{18}F-FCH in the detection of other tumours is not well studied and the results do not always show a higher sensitivity of ^{18}F-FCH compared to ^{18}F-FDG. Therefore new studies have to be performed to address the role of ^{18}F-FCH in other tumours.

2.1.1.7 PET IMAGING IN OSTEOSARCOMA

In OS, PET imaging is mainly used for monitoring of treatment response (Fig 2.2). In a study performed on 36 patients between 3 and 19 years of age, PET imaging with ^{18}F-FDG was performed before and after chemotherapy: the mean SUV after chemotherapy was significantly higher in the poor responder group compared to the good responder (p=0.04) and the cut-off value after treatment for a favourable response was < 2.5 (Kim, Kim et al. 2011).

Franzius et al. assessed that ^{18}F-FDG PET imaging may complement the other factors and tools in defining the prognosis in OS: he performed a study on 29 patients with OS and he found that high ^{18}F-FDG uptake, measured as a ratio between tumour and non-tumour ratios is prognostic of poor outcome (Franzius, Bielack et al. 2002).

Along the same lines Costelloe et al. claims that ^{18}F-FDG PET imaging can be used as a prognostic indicator of tumour response and of survival. The study was conducted on 31 patients with OS finding that SUVmax values > 15 were associated with a worse progression-free survival and 3.3 was kept as a cut-off for a worse prognosis in case of patients showing higher SUV (Costelloe, Macapinlac et al. 2009).

As for tumour grading some studies have been performed but Buck et al., on the other side, claim that ^{18}F-FLT correlates more than ^{18}F-FDG with the tumour but no difference is present in ^{18}F-FLT uptake between grade 2 and 3 tumours (Buck, Herrmann et al. 2008).

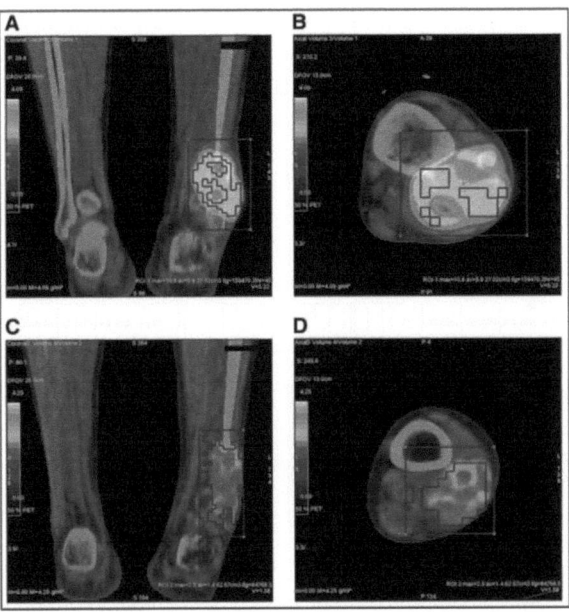

Fig.2.2: Example of ^{18}F-FDG PET/CT images from a 25 years old man with high-grade conventional OS in the metaphysis: coronal (A) and axial (B) images of tumour (arrows) before chemotherapy shows high 18F-FDG uptake in the tumour. After chemotherapy the coronal (C) and axial (D) images show a much reduced and intense ^{18}F-FDG uptake suggesting the treatment was effective (adapted from (Costelloe, Macapinlac et al. 2009).

Despite the promising results, there are still many limitations in these studies: first of all the low number of patients available and also the absence of long lasting follow up.

Further not many additional tracers have been tested in OS and so far combination of more tracers in PET imaging was never applied before treatment to characterise the biology of each individual tumour. This would be extremely helpful in designing a therapy that would target specifically the processes that are found to be mainly deregulated in each patient.

2.2 PHOTODYNAMIC THERAPY

2.2.1 BASIC PRINCIPLE

Photodynamic Therapy (PDT) is an old treatment modality which only recently got recognised and is applied for neoplastic and non-malignant diseases.
It consists of the combination of three different components that are individually not toxic: oxygen, photosensitizer (PS) and light.
After being injected in the whole body light of a specific wavelength activates the PS locally where the light is applied and then the activated PS is able to direct a series of phenomena that involve the vasculature in the tumour and the immune system (Fig. 2.3).

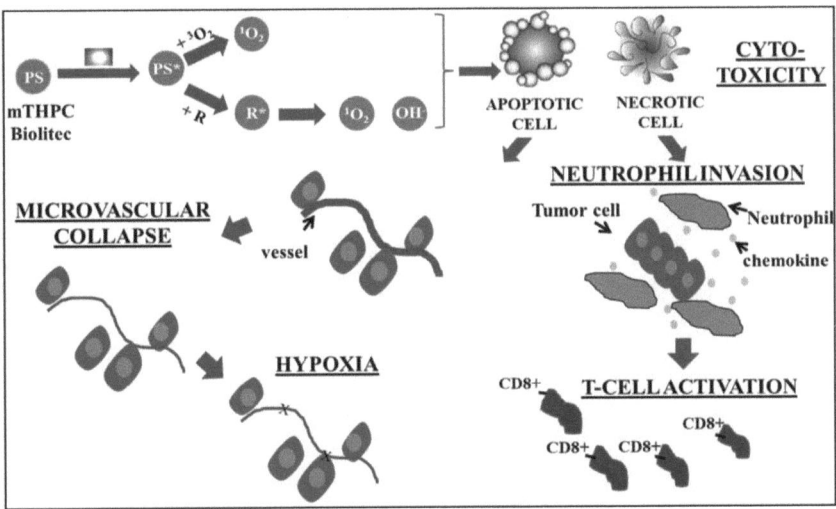

Fig.2.3: Schematic overview of the mechanism of action of Photodynamic therapy (adapted from (Castano, Mroz et al. 2006).

Similar to chemotherapeutics, also PDT is not only damaging the tumour cells but also can induce toxicity in normal tissue however compared to other chemotherapeutics, since the illumination and consequently the activation of the

PS is local, the side effects are much more limited. Furthermore none of the PS accumulates in the nuclei thus not inducing any carcinogenic consequence.

The local illumination represents also a disadvantage because the PDT is not effective against metastatic lesions which are the main cause of death in cancer patients as well as in OS patients (Agostinis, Berg et al. 2011).

2.2.2 PHOTOSENSITIZERS

Photosensitizers (PS) are molecules that need to be activated by a specific wavelength light in order to produce toxic substances and reactive oxygen species (ROS) that will damage the cells. Since they are captured by all the cells in the body it is important to define a specific illumination time when the accumulation in the tumour cells is greater than in normal tissues.

After the PS is in an excited state, electrons are transferred from PS to other molecules inducing consequently two reactions: the first one consists in transferring electrons and hydrogen atoms from the PS to a substrate which gets excited and since the excited state has a short lifetime, around ~3 µs, the PS can react with the substrate only if the substrate is max $2\text{-}4 \times 10^{-6}$ $cm^2 s^{-1}$ distant from the oxygen molecule. The second type of reaction involves transfer of energy from PS to the ground state oxygen ($^1O_2$) which consequently will give rise to the production of reactive oxygen intermediates which include also peroxides (Luksiene 2003; Josefsen and Boyle 2008; O'Connor, Gallagher et al. 2009).

The most used PSs in the clinic have a tetrapyrrole structure which is similar to the protoporphyrin present in the haemoglobin.

The main features of an "ideal" PS include: it should not be toxic or carcinogenic; its excitation light should be between 600 and 800 nm because above 800 nm there is not enough energy to excite the oxygen and produce radicals; it should not be darktoxic; it should accumulate preferentially in the tumour area and less in the normal tissue; it should be cleared up from the normal tissue to reduce the side effects; depending on the area of treatment it should induce a strong inflammatory response or not (in the case of the brain tumours); it should not photobleach though some recent findings show that this phenomenon is helpful to destroy the PS without having problems of overtreatment; it should be easily synthesised and the drug formulations should be stable over years (Allison, Downie et al. 2004; Josefsen and Boyle 2008).

The PSs are divided in two main groups depending on the chemical structure that they have: porphyrin or non-porphyrins. The first ones are further subdivided in first, second or third generation.

The first PS used in clinic was Photofrin® that comes from chemical modifications of the hematoporphyrin and it is composed of 60 compounds though it is unknown which of the 60 is the effective compound. It is considered at the moment the gold standard for PDT and it has been used for the treatment of various cutaneous lesions and benign or malignant tumours like lung, esophageal and bladder cancers as well as skin disease (including malignant tumours) and it is under investigation for the use in Kaposi's sarcoma, Barrett's esophagus with high-grade dysplasia, psoriasis, brain, breast, head and neck cancers (Josefsen and Boyle 2008; O'Connor, Gallagher et al. 2009).

Despite the successful results, the Photofrin® as well as all the PSs from the first generation display several disadvantages: unclear definition of the effective component among the 60 that are present in the Photofrin®, high injected dose (1.5-2 $mgkg^{-1}$), treatment start 48 hrs after its injection, long-lasting photosensitivity and consequently sunlight protection from the patients for at least 4 weeks and low wavelength of absorbance (630 nm) (O'Connor, Gallagher et al. 2009; Agostinis, Berg et al. 2011).

Therefore new compounds have been tested that are classified as the second generation photosensitizers among which Foscan®, Visudyne®, Tookad® and Levulan®.

All of them share the same advantages compared to the first generation group including a faster clearance time from the body, a faster biodistribution time before illumination, lower photosensitivity and consequently reduced side effects.

Among these the most potent one is Foscan® whose active substance is 5,10,15,20-tetrakis(meta-hydroxyphenyl)chlorine (mTHPC): it is at the moment used as a treatment in head and neck cancer and under investigation in gastric and pancreatic cancers and in other non-malignant pathologies. The main advantages are the low effective doses (0.1 mgkg^{-1}) and light intensity (10 Jcm^{-2}) though the phototoxicity is present up to 20 days which is better than in the first generation compounds but much worse compared to the Visudyne® which has only 1-2 days of photosensitivity. Furthermore the half-life in the blood is quite long (45-65 hrs).

Another promising PS is Tookad®, a palladium-metalated bacteriopherophorbide coming from a bacterial analogue of the plant chlorophyll. It gained a lot of interest since the absorption is maximum at 763 nm therefore the depth that the light reaches is 4mm, the plasma half-life is ~20 min and the compound is cleared from the circulation in mice 15-20 min after injection reducing the accumulation of this compound in the tissues. In human patients it was used so far only in 24 prostate cancer patients and at the early follow-up patients displayed not many side effects or complications. So far it is not yet approved in the clinics but the findings coming from the first clinical studies are encouraging.

A last example is represented by Levulan® which is composed of 5-aminolevulinic acid (ALA). This substance is also produced endogenously since it is a precursor of the porphyrin present in the heme biosynthetic pathway implying that the synthesis is quite easy as well as the formulation. It is mainly used for the treatment of superficial lesions like actinic keratosis or for superficial carcinomas since the absorption peak is at 410 nm with four smaller peaks at 510, 540, 580 and 635 nm.

Other improvements were achieved with the third generation compounds above all on the solubility and specificity of the PSs. For example researchers tried to combine the second generation PSs with antibody that would directly bind to tumour cells antigen or with other carriers like pegylated liposomes or nanoparticles. For the third generation compounds, studies are in a pre-clinical stage therefore still further tests have to be performed (O'Connor, Gallagher et al. 2009).

In this thesis we used two investigational liposomal formulations of Temoporfin (mTHPC), Formulation 1 and Formulation 2, provided by biolitec research GmbH. Formulation 1 was used for the *in vitro* experiments and Formulation 2 for the *in vivo* experiments. As the only difference between these two compounds is that Formulation 2 is dissolved in 5% glucose while Formulation 1 is dissolved in water, we hypothesise that the biological properties of the two compounds can be considered equal. Both formulations are still under investigation however in pre-clinical testing compared to Foscan® they showed both *in vitro* and *in vivo* more efficacy and faster distribution time. Therefore our choice was directed towards these new formulations.

Concerning the nonporphyrin compounds, there are not many on which the clinic focused on. Hypericin was, first, isolated in a plant from Hypericum species and whose activity is strongly dependent on the oxygen concentrations. It was shown to be effective in repressing the tumour growth in a RIF-1 (radiation-induced fibrosarcoma) mouse tumour model when the illumination was performed 30 min after intravenous injection of the PS. The same result was observed in a xenograft subcutaneous model of squamous cell carcinoma.

2.2.3 MECHANISMS OF PDT ACTION ON THE TUMOUR

There are three mechanisms of action of PDT in the tumour: cytotoxicity, vasculature contribution and immune system activation (Fig.2.3).

2.2.3.1 CYTOTOXICITY

Cytotoxicity starts already few minutes after irradiation when the active cellular transport is reduced, the membrane gets depolarised and the activity of the enzymes in the plasma membranes is inhibited. Plasma, mitochondrial and nuclear membranes get oxidised and even if the membranes integrity is still kept, the ionic homeostasis is lost and consequently all the ions are secreted from the cells which induces an acute inflammatory response. The increase in Ca^{2+} concentration triggers the release of arachidonic acid which binds and activates the phospholipase A_2 promoting the generation of superoxide radicals via the proteinkinase C. From this moment apoptotic signals are switched on as well as necrotic and autophaghy-related pathways (Luksiene 2003).

Caspases-mediated apoptosis is the most frequent cytotoxic phenomenon but also necrosis was shown to play a role once the mitochondria are damaged and when the Ca^{2+} concentration increases. Autophagy might also play a role in the PDT-caused cytotoxicity through the macroautophagies which usually are involved in the degradation or recycle of protein and organelles. Despite there are clear indications of contribution from all these cellular death pathways, it is not yet understood how these three mechanisms are interrelated.

Recently new studies investigated also the cytoprotective mechanisms which depend on the tumour cell type and which are most probably regulated by the expression of detoxifying enzymes and of heat shock proteins (HSPs) (Agostinis, Berg et al. 2011).

2.2.3.2 VASCULATURE CONTRIBUTION

After PDT the microvessels collapse occurs and consequently tumour cells become hypoxic (Fig.3.1). This phenomenon is strongly dependent on which PS is used. Photofrin® induces vascular leakage in consequence of which the activated platelets start to attract leukocytes in the damaged area where they contribute to the thrombus formation.

The main factor influencing the effect on the vasculature in each PS depends on the drug-light interval and on the chemical structure of the PS. For example Kurohane et al. showed that PDT with Verteporfin (Visudyne®) induced an inhibition in the tumour growth already 15min after irradiation through a severe damage to the endothelial cells more than directly inducing cytotoxicity (Kurohane, Tominaga et al. 2001).

Chen et al. used a mouse model of RIF (radiation induced fibrosarcoma) tumour to compare the effects of the treatment 15 min and 6 hrs after irradiation with intravenously injected hypericin used at a concentration of 5 mgkg^{-1}. In accordance with the previous findings, Chen et al. also found that the short drug-light interval was more effective (Chen, Pogue et al. 2003).

Concerning Foscan®, many studies suggest that there is no correlation between the time of maximal response of the tumour with the time of maximal uptake in the tumour. Indeed Cramers et al. treated mesothelioma xenograft mouse model using different drug-light intervals showing that the maximal PDT response was 1-3 hrs after irradiation of the PS. Despite these results, the protocol used nowadays suggests to irradiate the tumour 48-120 hrs after injection when the PS reaches the maximal uptake in the tumour (Cramers, Ruevekamp et al. 2003).

Same results were obtained by Jones et al. on a rat model of fibrosarcoma: 2hrs after injection with Foscan® they reached the highest serum concentration and 24hrs after PS injection they reached the maximal tumour concentration.

New studies are pointing to the possibility of an increase of vessel permeability after PDT which augmented the accumulation of Doxil ,a liposomal formulation of doxorubicin (Snyder, Greco et al. 2003). These findings are still under investigation and new aspects of this correlation between PDT and vasculature will be further evaluated.

2.2.3.3 IMMUNE SYSTEM ACTIVATION

The most important experiment that showed the influence of immune system on the PDT efficacy was carried out by Korbelik et al. (Korbelik, Krosl et al. 1996). They injected EMT6 mammary sarcoma cells subcutaneously in syngeneic BALB/cJ mice, in SCID and nude mice. The nude mice do not have a thymus so they lack T cells but they have normal production of natural killer (NK) and B lymphocytes. The SCID mice, on the other side, have a mutation in a gene encoding for a phosphatidylinositol kinase which is necessary for the recoupling of the DNA strands during the recombination V(D)J needed for the maturation of B and T lymphocytes. Therefore they do have normal myeloid and NK production but they lack mature B and T lymphocytes. They produce only a low percentage of $CD4^+$ lymphocytes which are from oligoclonal origin (Korbelik, Krosl et al. 1996).

The treatment was started when the tumour was 5-7 mm large and the thickness not more than 3.5mm. The PDT protocol consisted in injecting Photofrin® ($10 mgkg^{-1}$) and applying either 110 Jcm^{-2} or 220 Jcm^{-2}. The results showed that the treatment was effective in BALB/c and SCID mice but in the second group the tumour regrew already 2 weeks after PDT: even when a double energy dose was applied (220 Jcm^{-2}) the effect was delayed by approximately 5 days (Fig. 2.4).

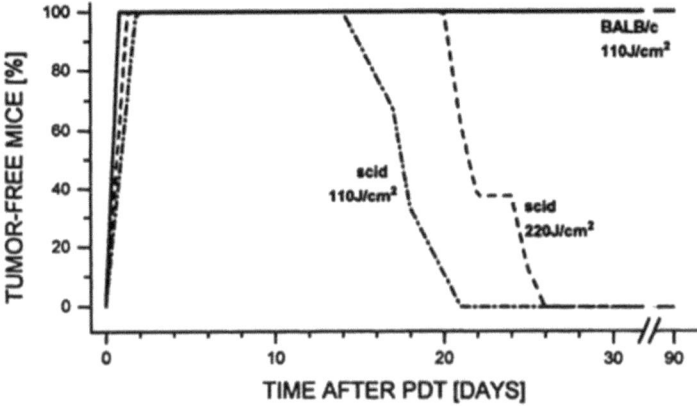

Fig.2.4: Treatment response in BALB/c and SCID mice after PDT performed with Photofrin® (10 mgkg^{-1} i.v.) and with either 110 Jcm^{-1} or 220 Jcm^{-1} energy dose (adapted from (Korbelik, Krosl et al. 1996).

Same results were observed with the nude mice (Fig. 2.5).

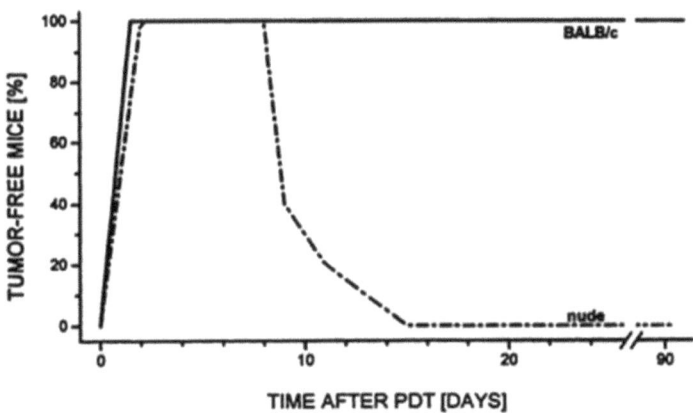

Fig.2.5: Treatment response in BALB/c and nude mice after PDT performed with Photofrin (10 mgkg^{-1} i.v.) and with 110 Jcm^{-1} energy dose (adapted from (Korbelik, Krosl et al. 1996).

After these findings they tried to restore the T-cells presence which is almost abolished in the SCID mice by transferring the spleen from BALB/c mice but they did not get an improvement in the PDT efficacy. Therefore they decided to transplant bone marrow from BALB/c to either 3 weeks-old SCID mice or 7 weeks-old SCID mice, and they also transplanted bone marrow from SCID to BALB/c mice and from SCID to SCID mice as a further negative control. They could show that the transplant of bone marrow from BALB/c mice restored the treatment efficacy in the SCID mice reaching almost 100% of cure (Fig. 2.6).

Further they could show that the SCID mice that received the bone marrow transplantation when they were 3 weeks old responded to the treatment better (80% of mice cured) than the ones that received the bone marrow transplantation when they were 7 weeks old (63% of mice cured). Along with these results, the BALB/c mice, with a bone marrow received by a SCID mouse, show recurrence of the tumour 14 days after PDT as well as the SCID mice used as a negative control that got the bone marrow from another SCID mouse.

In summary this experiment points to the contribution of the T-lymphocytes to the PDT efficacy and to the importance of these cells in order to have a long-lasting effect of the treatment more than to the initial induction of tumour cells death (Korbelik, Krosl et al. 1996).

These findings led the researchers to focus on the contribution on the innate and adaptive immune system after PDT and nowadays it is thought that there are two different steps in the effect of PDT mediated by the immune system.

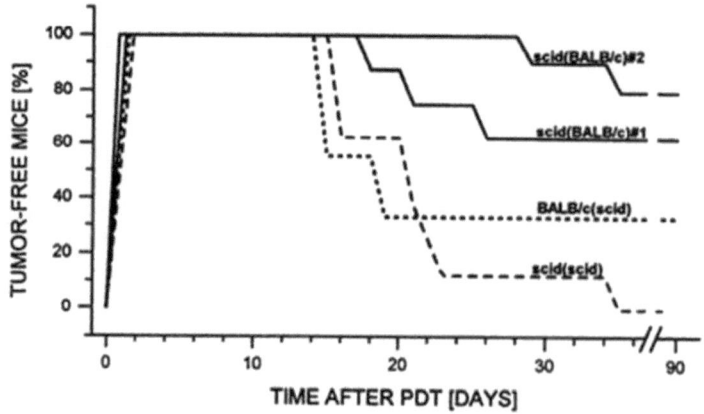

Fig.2.6: Treatment response in BALB/c and SCID mice after PDT performed with Photofrin (10mgkg^{-1} i.v.) and with 110 Jcm^{-1} energy dose. The SCID mice received the bone marrow from the BALB/c mice when they were 3 weeks (SCID (BALB/c)#2) or 7 weeks old (SCID (BALB/c)#1). BALB/c mice received the bone marrow from SCID mice (BALB/c (SCID)) and as a negative control SCID mice got the bone marrow from another SCID mouse (SCID(SCID)) (adapted from (Korbelik, Krosl et al. 1996)).

First of all the innate immune system, mainly represented by neutrophils, is activated towards the tumour cells. The neutrophils, through the release of several cytokines, attract the monocytes and immature dendritic cells that are then activated via a cell-to-cell contact and TNF-α secretion. It has been found that serum of tumour bearing mice after PDT contained high level of IL-6, macrophage inflammatory protein 1 and 2, prostaglandins secreted by endothelial cells, leukocytes and tumour cells (Castano, Mroz et al. 2006; Kousis, Henderson et al. 2007).

This last step is the beginning of the second part of the immune system activation which involves the adaptive immune system: T-cells infiltrate the tumour area, differentiate and secrete INF-γ (Kousis, Henderson et al. 2007). This flow of subsequent processes is quite similar to the response of the immune system to infections.

Therefore despite the PDT is a local treatment, the effects can be systemic in light of the activation of the immune system.

Since the immune response has such an important role for the PDT efficacy, recent studies tried to evaluate the possibility to combine the PDT with other therapies.

One strategy is to stimulate the innate immune system using microbial stimulators of the Toll-like receptors (TLRs) that are present on the membrane of monocytes, macrophages, dendritic cells, mast cells and some epithelial cells. TLRs are able to trigger the NFκB signalling and activate, consequently, the innate immune system.

Based on this principle, Myers et al. performed PDT before and after administration of Corynebacterium parvum (CP) in a mouse model of subcutaneous bladder cancer and they found that when CP was given after PDT the efficacy was significantly higher compared to a PDT pre-treatment with CP (Myers, Lau et al. 1989).

In another study, the Bacille-Calmette Guérin vaccination (BCG) was applied in mice with EMT6 subcutaneous tumour and PDT was performed using 6 different PSs: an increase was observed in the number of cured tumours and of memory T-cells in the group of mice that received the BCG compared with PDT alone (Korbelik, Sun et al. 2001).

A similar result was achieved on mice with NRS1 squamous cell carcinomas. These mice were injected intratumorally 3 hrs before PDT with OK432, preparation containing killed streptococcal bacteria, and they showed an increase in tumour-free time compared to the mice that did not get OK432 or to the ones that got OK432 only after PDT (Uehara, Sano et al. 2000).

A second strategy to increase the photodynamic therapy efficacy involves the administration of cytokines like TNFα as Bellnier et al. showed. Three hrs before injection of different concentrations of Photofrin® and energy dose (2.5 mgkg^{-1} and 288 Jcm^{-2} or 5 mgkg^{-1} and 144 Jcm^{-2}) they administered recombinant human TNFα (rHu-TNFα) and they got a 100% response from the mice from this group compared to a response of 81% in the case of mice injected with 2.5 mgkg^{-1} of Photofrin® and 288 Jcm^{-2} energy dose and the rHu-TNFα just before irradiation. In the mice treated only with rHu-TNFα the percentage of success was even lower (20%). In the case of mice treated twice with 5 mgkg^{-1} and 144 Jcm^{-2} the percentage of mice with a tumour regression was 100% suggesting that the rHu-TNFα gives an additive synergistic effect (Bellnier 1991).

The third and last approach aims at increasing the anti-tumour response by combining the administration of dendritic cells with PDT. It was shown that dendritic cells when exposed *ex vitro* to tumour antigens and re-administered have the ability to induce tumour regression in several mouse models of cancer.

Along this principle, Saji et al. tried to directly inject the dendritic cells in the tumour using ATX-S10 Na(II) as PS to treat BALB/C mice with two different tumours: colon (CT26) and melanoma (B16). Dendritic cells were administered the day after irradiation and then again at day 14, 15 and 17 and they got a statistically significant higher ($p<0.0006$) overall survival compared to the other groups that received the single treatment. Further the mice that received the treatment with dendritic cells on one tumour could show a regression also on the other non-treated tumour (Saji, Song et al. 2006).

A last aspect to mention concerns the immunosuppressive effect that PDT has when IL-10 is produced and this occurs mainly when the PDT is performed using red light and to a much lower extent when the treatment involves only the skin (Castano, Mroz et al. 2006).

2.2.4 PDT IN THE CLINICS

In clinic practice PDT has been used from 1970s applying hematoporphyrin derivative (HPD) in 5 patients with bladder cancer. Consequently around 200 clinical trials were initiated and they include many and different malignant and non-malignant diseases. Here below you see the list of the PSs used in the clinic already with reference to the cancer application where the PS is being used (Table 2.3).

PHOTOSENSITIZER	STRUCTURE	WAVELENGHT (ABSORPTION)	APPROVED IN	TRIALS IN	CANCER APPLICATION
Porfimer sodium (HPD)	Porphyrin	630nm	Worldwide		Lung, esophagus, bile duct, bladder, brain, ovarian
ALA	Porphyrin precursor	635nm	Worldwide		Skin, bladder, brain, esophagus
ALA esters	Porphyrin precursor	635nm	Europe		Skin, bladder
Temoporfin	Chlorine	652nm	Europe	United States	Head and neck, lung, brain, skin, bile duct
Verteporfin	Chlorine	690nm	Worldwide	United Kingdom	Ophthalmic, pancreatic, skin
HPHH	Chlorine	665nm		United States	Head and neck, esophagus, lung
Motexafin lutetium	Texaphyrin	732nm		United States	Breast

Table 2.3: List of the PSs used in the clinic

For example in skin cancer ALA and its derivative are approved in Europe, USA and Canada for the treatment of actinic keratosis as well as in Europe and Canada, these PSs are used for the treatment of basal cell carcinoma. On squamous cell carcinoma (SCC) of the skin many studies have been performed but the recurrence rates were higher than 50% (Agostinis, Berg et al. 2011).

Another example is the use of PDT in head and neck tumours: a small randomized clinical trial was performed to compare the efficacy of treatment performed with cisplatin and PDT using 5-FU or with PDT using porfimer sodium. Patients were showing mainly SCC of the cavity, pharynx or larynx, Kaposi sarcoma and SCC of head and neck. The PDT gave a better clinical response. With the development of second generation PSs, new clinical trials started where they tested temoporfin for the treatment of oropharyngeal cancers and they found that 85% of the patients (97 out of 114) showed a complete response 12 weeks after treatment with a disease free survival rate of 75% at 2 years.

In another study PDT was used on patients with stage I, II or III of second or multiple primary head and neck tumours treated with temoporfin. They obtained a cure rate of 85% in stage I tumours and 38% in stage II or III.

An additional clinical trial was performed on 128 patients with advanced head and neck cancer. The patients enrolled for this trial did not have response after the conventional treatment or were not suitable for it. PDT was performed 96hrs after Temoporfin injection and in 43% of patients the tumour completely regressed while in the rest the tumour volume was reduced of around 50% (Agostinis, Berg et al. 2011).

In prostate cancer the main limitation of surgery and ionising radiotherapy is the high morbidity because of close presence of rectum, nerves and bladder.

One small study has been performed on patients with recurrent prostate cancer using motexafin lutetium (MLu). Beside some mild side effects related to urinary toxicity, the study showed an improvement in the biochemical and pathological disease after use of PDT.

Another study focused on the use of palladium (Pd)-bacteriopheophorbide (padoporfin) PS to treat patients affected by prostate adenocarcinoma with a local failure after radiotherapy. In a phase I clinical trial, 24 patients were treated with this PS without any complications therefore they further treated 28 patients in a follow up studies with increasing energy doses concluding that the tumour response correlated with the energy dose used (Agostinis, Berg et al. 2011).

The superficial types of bladder cancer are relatively suitable for PDT. Some studies investigated the possibility of using HPD for the treatment of 50 superficial bladder transitional cell carcinomas (TCCs) in 37 patients reaching a 74% complete response.

In another study they used HPD-PDT in 34 patients with bladder carcinoma in situ and they showed a 73.5% of complete response 3 months after treatment but 2 years after 77.8% of the patients developed a recurrence.

Recent studies further suggested that in this tumour the combination of PDT with multiple BCG treatment or with other chemotherapies may augment the PDT efficacy for this cancer (Agostinis, Berg et al. 2011).

2.2.5 PDT IN OSTEOSARCOMA

Not many studies have been published so far about the use of PDT for the treatment of OS. One animal study was performed in dogs which develop spontaneously OS with high percentage above all when they have long rear extremities. In this experiment they applied 0.4 mgkg^{-1} of Vetoporfin (Visudyne®) intravenously in 7 dogs with OS and used 500Jcm of energy dose. One dog was used as negative control in which only the energy dose equal to 500Jcm was applied (Burch, London et al. 2009).

The main purpose of this pilot study was to evaluate whether PDT could cause necrosis in large osseous tumours and to calculate the volume of treated tissue. The necrosis was clearly visualised via H&E staining on the tumour sections and it correlates with the necrotic signal visible on the MRI. Since OS in dogs resemble the one in humans, it can give also more general information about the efficacy of OS in humans (Burch, London et al. 2009).

The results shown by Burch et al. are in agreement with the results previously published by other groups that used subcutaneous OS mouse models (Kusuzaki, Minami et al. 2000; Nomura, Yanase et al. 2004).

Despite the few published data on the efficacy of PDT *in vivo* there are many limitations in these studies to be considered: in the case of Buch et al. low number of animals, the need of amputation of the leg 48hrs after PDT for ethical reasons, no follow-up information about the effect of PDT on the dog's survival, high heterogeneity in tumour volume among the dogs; in the case of Kususzaki and Nomura et al. the use of subcutaneous model which does not reflect the tumour microenvironment of the bone.

2.2.6 APPLICATION ROUTES FOR PDT

The possibilities of application routes in PDT are different: intratumoural, intraoperatively and endobronchially. The first one is shown by Buch et al. and consists in the intravenous injection of the PS followed by a local irradiation with a specific light through a 0.94mm fiber optic cable that was localised in the tumour via a 2mm steel guide rod that was placed through an incision over the proximal anterior surface of the affected radius (Fig. 2.7) (Burch, London et al. 2009).

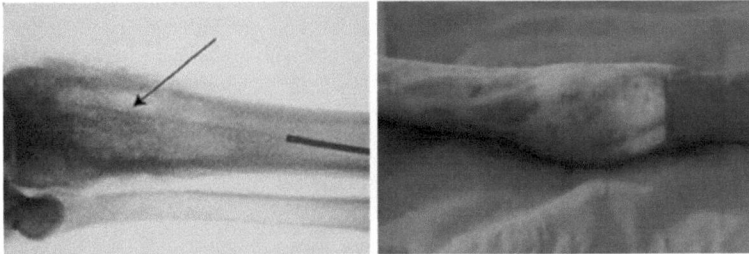

Fig.2.7: Intratumour application of PDT in a dog with a spontaneous OS using a fiber optic cable (arrow in left picture) and application of the light (right picture) (adapted from (Burch, London et al. 2009)).

Another application is intraoperative: the tumour cells can be visualised taking advantage of the fluorescence quality of the PS. In this way the medical doctors can be guided in the resection and after the surgery local irradiation can be applied (Fig. 2.8) (Zimmermann, Ritsch-Marte et al. 2001).

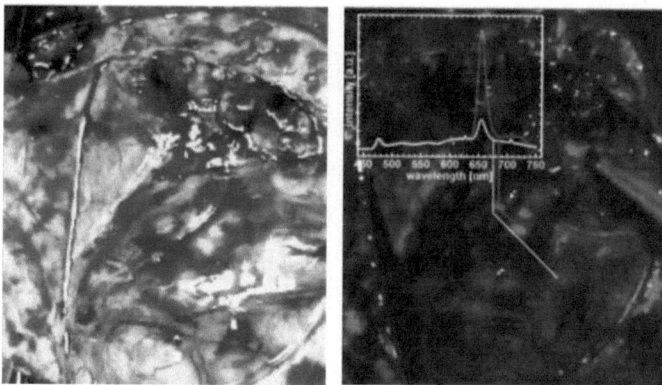

Fig.2.8: Intraoperative application of PDT on a human patient with malignant glioma: tumour visualised with a normal white light (left picture) and with a blue-light fluorescence with the tumour cells in red (right picture) (adapted from (Zimmermann, Ritsch-Marte et al. 2001).

The third and last possibility is the endobronchial application for PDT that can be used as a palliative treatment of endobronchial metastases that can significantly reduce airway obstruction caused by metastases (Fig. 2.9) (Litle, Christie et al. 2003).

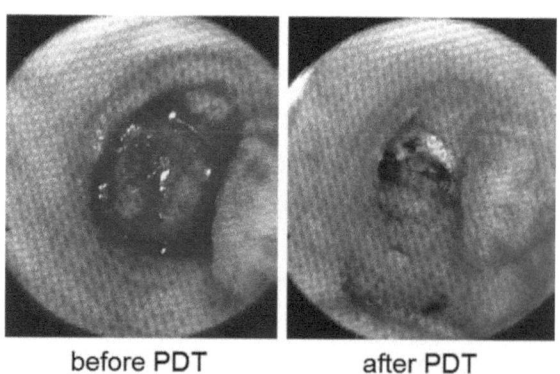

before PDT   after PDT

Fig.2.9: Endobronchial application of PDT on a human patient with a breast endobronchial metastasis: pictures of the affected area before and after PDT (adapted from (Litle, Christie et al. 2003)).

2.3 AIMS OF THE THESIS:

In this thesis we investigated on one side the potential of PET imaging in characterising different OS phenotypes and detect metastases in the lung; on the other side we evaluated the efficacy of PDT for OS treatment. Both aims are focused on the need of improving the survival of patients with metastases by assessing an additional role of PET imaging in the design of a therapy and by evaluating new therapeutic strategies.

For the first aim we selected three tracers (^{18}F-FDG; ^{18}F-FMISO; ^{18}F-Fluoride) and compared the uptake of these tracers in the primary tumour and the control leg, used as our reference, in three orthotopic, highly metastasising mouse models of OS that develop different radiological phenotypes.

The other three tracers (^{18}F-FLT, ^{18}F-FET and ^{18}F-FCH) that were also included in this study did not show consistent results in one OS mouse model. Each of the three tracers, used in this study, targets a specific metabolic process thus allowing us to define the specific features of each tumour model based on changes in these processes: ^{18}F-FDG is indicator of glucose metabolism; ^{18}F-FMISO accumulates in hypoxic regions and ^{18}F-Fluoride is an indicator of bone remodelling. The hypothesis behind is that depending on the different pattern of tracer uptake for each tumour, a specific therapy strategy could be designed: this treatment protocol should target the biological processes altered in this individual tumour.

Additionally we wanted also to evaluate the possibility of detecting lung metastases with one of these three tracers that were highly sensitive in the detection of the primary tumour.

In the second aim of my thesis we investigated the efficacy of PDT for the treatment of OS.

To fulfil this aim we performed studies *in vitro* and *in vivo* using non-PEGylated liposomal formulations Temoporfin: Formulation 1 (biolitec research GmbH,

Jena-Germany) for the *in vitro* experiments and a newly more stable non-PEGylated liposomal formulation, namely Formulation 2, for the *in vivo* part of the project.

In order to overcome the limitations from the published data on mice where they used only subcutaneous models we established an orthotopic OS mouse model where OS cells are injected intratibially.

The sub aims of the *in vitro* and *in vivo* parts are listed below:

In vitro:

1. evaluation of the uptake of the Formulation 1 in different OS human cell lines
2. measurement of the cytotoxic effects of PDT on one human OS cell line chosen for the *in vivo* studies, namely 143B.

The *in vitro* results were published recently by our group and part of the material & methods and results were taken from the published paper and further extended where more experiments were performed compared to the published ones (Reidy, Campanile et al. 2012).

In vivo:

1. measurement of Formulation 2 uptake in the tumour, lung and liver using a specific optic fiber provided by biolitec research GmbH (Jena-Germany) on a xenograft and a syngeneic mouse models
2. optimisation of the protocol for PDT *in vivo* using a xenograft model of OS

2.3.1 OS MOUSE MODELS

An ideal OS mouse model reflects the human disease displaying similar pattern of markers' expression, metastatic spread, aetiology and treatment response. Further a model that would produce the tumour spontaneously or after induction with an oncogene or a chemical substance reflects more the human disease. A good model of OS is the dog that develops the tumour spontaneously.

In mice the rate of spontaneous OS is really low: the first OS mouse models was obtained in 1938 by Brunschwig et al. who injected intratibially 1,2-benzpyrene and cholesterol. Another model was generated between the 1960s and 1970s after exposure to radiation, radioactive isotopes and oncogenic viruses.

Later in the 1980s five clones from a spontaneous Balb/C mouse OS were isolated and characterised *in vitro*.

Nowadays most of the OS mouse models are generated from the injection of tumour cells orthotopically or subcutaneously in the mice. The intravenous injection is mainly used for the study on the pattern of migration and on the metastatic spread. Moreover the mouse model can be syngeneic or xenograft depending on the influence and importance of the immune system in the planned experiment.

Concerning the syngeneic models there are two cell systems: the first one consists of two clonally related cell lines K12 and K7M2 that show a different primary tumour and metastases development in vivo; the second model includes the low metastatic parental Dunn and highly metastatic derivative LM8 cell lines.

Concerning the first cell line system Khanna et al found that when injected intravenously 100% of the mice injected with K7M2 died within 17 days because of metastases development while only 20% of the mice injected with K12 died only 102.5 days after tumour cell injection. When tumour cells were orthotopically injected the rate of primary tumour development was > 90% in K7M2 tumour bearing mice compared to the 33% in K12 tumour bearing mice and the K7M2 tumour grew much faster than the K12. These differences in tumour growth rate are depending on a diverse pattern of deregulated genes as shown by Khanna et al. In K7M2 derived Khanna et al found a much higher expression of integrin β4, ezrin, clusterin, decorin and cerulopasmin which are influencing migration, adhesion and angiogenesis (Khanna, Prehn et al. 2000; Khanna, Khan et al. 2001).

The second syngeneic model consists of low metastatic Dunn and highly metastatic LM8 cells which derive from subsequent isolation of metastatic clones of Dunn cancer cells. The LM8 model is the most commonly used for testing the power of anticancer drugs, compounds coming from biological sources and radiation (Ek, Dass et al. 2006).

Our group showed that Dunn cells injected subcutaneously develop primary tumour and micro-metastases (≤0.1mm in diameter) mainly in the lung. The identification of micro-metastases was possible thanks to the *LacZ* tagging of the cells which allows visualising tumour cells after X-Gal staining of the perfused lung. LM8 cells were metastasising in the lung and liver in the form of macro- (> 0.1mm in diameter) and micro-metastases (<0.1mm in diameter) while Dunn tumours were producing only micro-metastases in the lung and in the liver (Arlt, Banke et al. 2011).

Orthotopic LM8 tumours are not the ideal model because they metastasise in ovaries, liver and kidney and to a lower extent in the lung (data not published) while Dunn tumours behaviour in the orthotopic location has to be further investigated.

The main advantage of the syngeneic models lay on the presence of the immune system that reflect more closely the situation in the patients where the immune system is actively influencing the tumour growth. On the other side the mouse development of primary tumour can be different for the molecular mechanisms, for the location of the metastases and this makes the syngeneic model not suitable for all kind of studies.

Additionally to this there are several human OS tumours. The main advantage in this case is the possibility to inject directly human OS cells that mainly metastasise exclusively to the lungs similarly to the human situation and they more closely would generate a microenvironment that could be similar to the one in the humans.

Among the human cell lines the most commonly used ones are U2OS, HOS, 143B, SaOS-2 and LM6.

U2OS are OS cells isolated from a 15 years old Caucasian girl with a moderately differentiated OS. These cells have been well characterised *in vitro* but *in vivo* they did not produce lung metastases if subcutaneously injected and the primary tumour developed in 63%. When these cells were intravenously injected, metastatic nodules developed in the lung in 100% of the mice after 8 weeks from tumour cell injection but when the cells were transfected with alkaline phosphatase (ALP) in both injection sites the primary tumour growth and the metastases formation were much lower. The orthotopic model of this cell line did not give clear results concerning the development of spontaneous lung metastases (Manara, Baldini et al. 2000).

HOS cells were isolated from a 13 years old Caucasian girl and they are also well characterised *in vitro* but *in vivo* so far no one successfully obtained primary tumour after subcutaneous or orthotopic implantation of these cells. Primary tumour growth and metastatic formation was reproduced, though, after *in vitro* transformation of the HOS with the oncogene v-Ki-ras. From this transformation two different cell lines were generated: KRIB and 143B. Both of them are able to form primary tumour and metastases after orthotopic injection within 4-6 weeks. The main limitation of both models is that they are transformed cell lines which do not truly represent the human situation since the acquisition of Ki-Ras mutation or amplification during tumour development is not so frequent in OS patients (Kido, Tsujiuchi et al. 1997). Consequently the Ki-Ras oncogene transformation might indeed affect pathway and molecular mechanisms that do not occur in the humans (Ek, Dass et al. 2006).

Finally SaOS-2 cells were isolated from an 11-year old Caucasian girl. When orthotopically injected they develop primary tumour in 2-5 months and lung metastases (Sabile, Arlt et al. 2011). A highly metastatic model of the SaOS-2 cell line was generated injecting the cells intravenously and 6 months after tumour cell injection pulmonary metastatic clones were isolated *in vivo* and re-injected intravenously. This process was repeated 5 times obtaining the LM6 that were highly metastatic, if intravenously injected.

In our laboratory we make use of most of the xenograft and syngeneic orthotopic OS mouse models described above.

To fulfill the first aim of my thesis we used three models: 143B, LM8 and SaOS-2. We chose these three models because they display extreme opposite radiological phenotypes (Yuan, Ossendorf et al. 2009) the primary tumour in the xenograft 143B model is characterised by the presence of lytic lesions in the tibia that start to be visible already in the second week of tumour development (Fig. 2.10-143B). Metastatic foci are formed mainly in the lung. The syngeneic LM8 model is slightly osteoblastic and it forms bony structures around the tibia that are visualised by x-ray from the third week after tumour cell injection (Fig. 2.10-LM8). Metastases are visible in the ovaries, liver, kidneys and lungs.

As for the xenograft SaOS-2 model, for my purposes we used a transformed cell line which was produced in our laboratory by Dr. Sabile, namely SaOS-2/Caprin-1, which has been characterised *in vitro* and *in vivo*. *In vivo* the SaOS-2/Caprin-1 model shows the same osteoblastic phenotype by appearance as the SaOS-2 model but the primary tumour grows much quicker: within 2 months primary tumour is detectable by x-ray in some mice while for the SaOS-2 model usually it can take from 2 to 5 months (Sabile, Arlt et al. 2013). The high heterogeneity in the primary tumour growth in the SaOS-2/Caprin-1 model forced us to perform the PET imaging in two different steps depending on the primary tumour development. In the Fig. 2.10 representative x ray-pictures from two different SaOS-2/Caprin-1 mice are visible: one mouse displayed detectable primary tumour already at day 43 after tumour cell injection while the other one showed typical osteoblastic lesions starting from day 85. Metastatic nodules are visible mainly in the lung.

Despite this heterogeneity the SaOS-2/Caprin-1 model grows faster and allows performing the experiment in much shorter time point. Another reason to choose these three models relates with the time of tumour progression: 143B and LM8 models endpoint is around 3-4 weeks though 143B tumour is characterised mainly by lytic lesion and the LM8 is mildly osteoblastic. The SaOS-2/Caprin-1 model is much slower and therefore it was interesting for us to compare the biology of fast and slow developing models by PET imaging.

Besides the orthotopic models, we made use also of a subcutaneous model where murine LM8 cells were injected in the right flank of C3H mice. Subcutaneous tumour was developing from the first week after tumour cell injection and metastases were developing mainly in the lung and in the liver. Mice were sacrificed 23 days after tumour cell injection and the use of this model relates mainly to the different site of tumour cell injection. To better identify the location of the primary tumour and to visualise the metastases to the level of a single cell, we make use in our laboratory of the *LacZ* system. Cells are *LacZ* tagged via a retroviral transduction and this enables us to visualise *ex vivo* the tumour cells after X-Gal staining (Arlt, Banke et al. 2011). Only one cell line (SaOS-2/Caprin 1) was not tagged with *LacZ* which was dependent on the earlier production of this cell line compared to the decision of using the *LacZ* tagging for our experiments but this does not affect our results since we did not count the number of metastases in the PET experiments or use the *LacZ* tagging for further analysis.

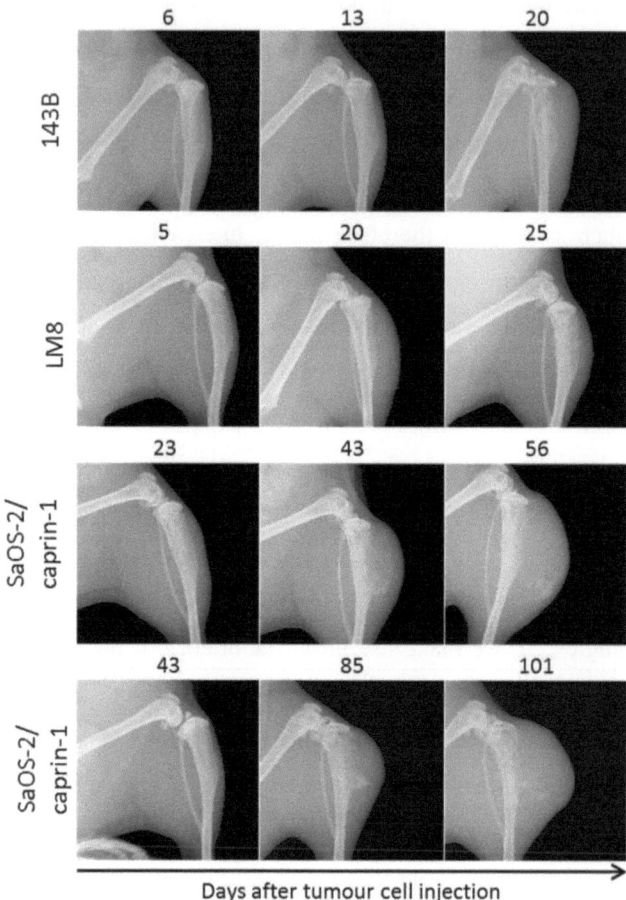

Fig.2.10: Representative x-ray pictures of mice from each OS mouse model during the primary tumour development.

CHAPTER 3

EVALUATION OF PET IMAGING IN OS MOUSE MODELS

Contributions
Study design: Carmen Campanile, Prof. Bruno Fuchs, Prof. Roger Schibli, Prof. Walter Born and Prof. Simon M Ametamey
Experimental design: Carmen Campanile, Dr. Matthias JE Arlt and Dr. Michael Honer
Experiment conduct: Carmen Campanile
Experiment planning: Josefine Bertz and Claudia Keller
Data analysis: Carmen Campanile
Analysis evaluation: Dr. Stefanie Krämer and Dr. Michael Honer
Animal project leader: Dr. Knut Husmann and Dr. Stefanie D Krämer
Animal responsible: Dr. Adrienne Müller
Animal work: Carmen Campanile, Dr. Matthias JE Arlt, Dr. Patrick Brennecke and Dr. Ana Gvozdenovic
Technical support: Josefine Bertz, Christopher Bühler, Claudia Keller and Petra Wirth
Material provider: Dr. Adam A Sabile, Cindy R Fischer and Matthias Nobs

3.1 RESULTS

3.1.1 UPTAKE OF ^{18}F-FET, ^{18}F-FLT AND ^{18}F-FCH IN THE 143B MODEL

The uptake of ^{18}F-FET, ^{18}F-FLT and ^{18}F-FCH was analysed in the 143B model to evaluate the sensitivity of these tracers in primary tumour detection.

^{18}F-FET which targets tumour cells with high rate of protein synthesis shows an ambiguous result: 2 mice show higer uptake in the tumour leg (Fig. 3.1A) and 2 mice have lower uptake (Fig. 3.1B) compared to the control leg which in the end resulted in an equal standardised uptake value for both legs when we performed the quantitative analysis (Fig. 3.1C).

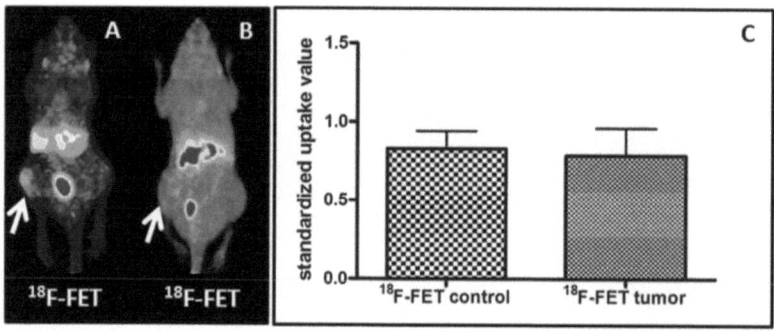

Fig.3.1: Summary of the ^{18}F-FET uptake in the 143B model: two representative PET coronal images as maximal intensity projection of OS bearing mice (tumour is in the left tibia-white arrow) injected with ^{18}F-FET (5A; 5B); quantitative analysis comparing the standardised uptake value in the control and tumour leg (5C) n=4. Data represent means +/- SEM (n ≥ 3).

^{18}F-FCH, which is specific for membrane turnover, shows a similar pattern of uptake compared to ^{18}F-FET since 2 mice show a higher uptake (Fig. 3.2A) and 2 mice show a lower uptake (Fig. 3.2B) in the primary tumour leg compared to the control leg. The quantification of the uptake in the two legs confirms the visual analysis: indeed no difference is present between the two legs in the total of 4 mice (Fig. 3.2C).

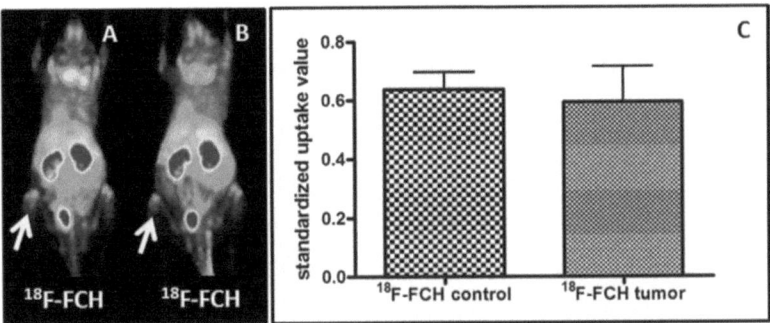

Fig.3.2: Summary of the ^{18}F-FCH uptake in the 143B model: two representative PET coronal images of OS bearing mice (tumour is in the left tibia-white arrow) injected with ^{18}F-FCH (6A; 6B); quantitative analysis comparing the standardised uptake value in the control and tumour leg (6C) n=4. Data represent means +/- SEM (n ≥ 3).

Finally ^{18}F-FLT is a tracer which marks the area of the tumour which are highly proliferative. Interestingly ^{18}F-FLT does not show any difference in tracer uptake between the two legs on the PET images (Fig. 3.3A-B) but the quantitative analysis demonstrates that the tumour leg displays a tendency to a lower uptake compared to the control which does not make sense considering the high tumour growth rate of this model (Fig. 3.3C).

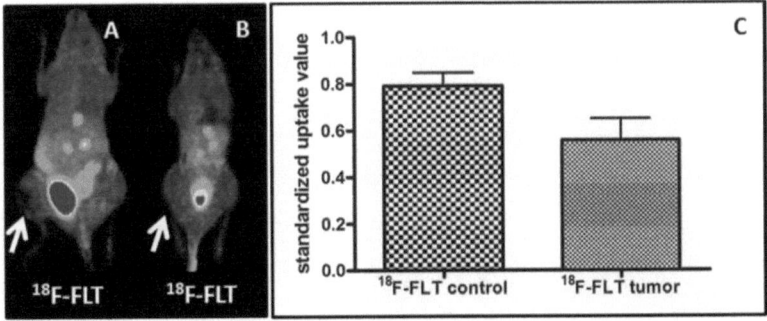

Fig.3.3: Summary of the ^{18}F-FLT uptake in the 143B model: two representative PET coronal images of OS bearing mice (tumour is in the left tibia-white arrow) injected with ^{18}F-FLT (8A; 8B); quantitative analysis comparing the standardised uptake value in the control and tumour leg (8C) n=4. Data represent means +/- SEM (n ≥ 3).

Since the results with these three tracers were not conclusive already with one model we decided not to further proceed with these three tracers and to focus only on ^{18}F-FDG, ^{18}F-FMISO, ^{18}F-Fluoride.

3.1.2 ^{18}F-FDG UPTAKE IN THE THREE OS MOUSE MODELS

^{18}F-FDG uptake, indicating high glucose metabolism, was detectable in the primary tumours of all three OS mouse models and defines the margins of the tumour (Fig. 3.4A). The mean SUVs of tumour and reference regions are shown in Table 3.1. The osteolytic 143B tumours showed the highest ^{18}F-FDG uptake with a 3.79±0.67 fold higher accumulation in the tumour than the reference region of the opposite leg. The LM8 model had 2.22±0.39 fold higher tracer uptake in the tumour than in the hip region and the SaOS-2/Caprin-1 model had a ratio of 2.04±0.15 between tumour and control SUVs. The difference in the ratios is statistically significant (p<0.05) between 143B and SaOS-2/Caprin-1 models (Fig. 9B). The sensitivity for ^{18}F-FDG was higher in 143B and SaOS-2/Caprin-1 models than in the LM8 model (Table 3.2). However, it should be noted that a different protocol for ^{18}F-FDG PET was used in the LM8 model.

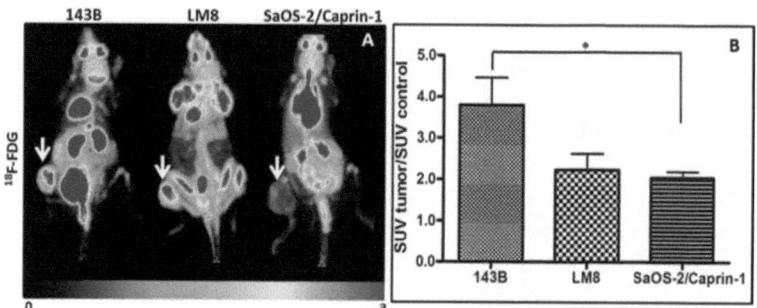

Fig.3.4: Representative PET coronal images of mice from the three different models injected with ^{18}F-FDG; the coloured pictures have a fixed SUV value for all the three pictures from a minimum value of 0 (black) to a maximum value of 3 (red) (A); quantitative analysis shows the ratio between the SUV in the tumour leg and in the control reference which is the other leg in the case of 143B and SaOS-2/Caprin-1 and the hip in the case of the LM8. Data represent means +/- SEM (n ≥ 3). *<0.05

Mouse Model	FDG-SUV	FMISO-SUV	Fluoride-SUV
143B control	0.357±0.08	0.156±0.008	2.04±0.25
143B	1.355±0.24	0.309±0.023	2.22±0.26
LM8 control	0.397±0.06	0.142±0.006	1.55±0.18
LM8	0.884±0.15	0.168±0.01	2.04±0.23
SaOS-2/Caprin-1 control	0.293±0.02	0.142±0.02	0.951±0.22
SaOS-2/ Caprin-1	0.598±0.04	0.318±0.03	2.166±0.37

Table 3.1: Mean±SEM values of tumour and reference SUVs for each tracer and each model

Tracer	Mouse models	Sensitivity (%)	No. of mice with detectable tumors/ Total No. of mice
^{18}F- FDG	Osteolytic (143B)	100%	4/4
^{18}F- FDG	Osteoblastic (LM8)	50%	3/6
^{18}F- FDG	Osteoblastic (SaOS-2/Caprin-1)	100%	5/5

Table 3.2: Sensitivity of PET imaging in detection of primary tumour with the ^{18}F-FDG

3.1.3 ^{18}F-FMISO UPTAKE IN THE THREE OS MOUSE MODELS

^{18}F-FMISO, an indicator of hypoxia, accumulated significantly in 143B and SaOS-2/Caprin-1 tumours but to a lower extent in LM8 tumours (Fig. 3.5A). SUV ratios were 1.99±0.15 for 143B and 2.23±0.24 for SaOS-2/Caprin-1 models (Table 3.1). Both were significantly higher (p<0.01) than the respective ratio in the LM8 model (1.18±0.08) (Fig. 3.5B). The sensitivity was good for all three tracers (Table 3.3).

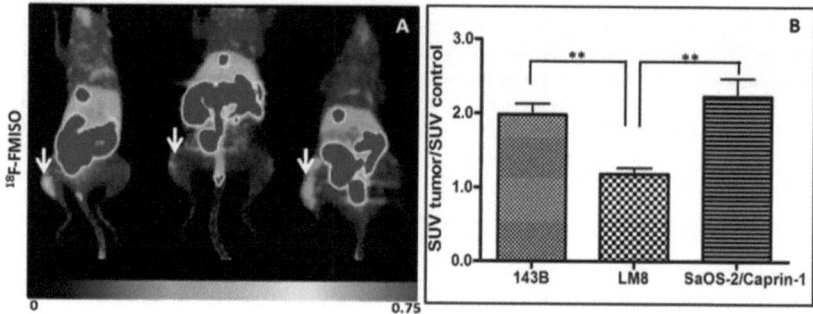

Fig.3.5: Representative PET coronal images of mice from the three different models injected with ^{18}F-FMISO; the coloured pictures have a fixed SUV value for all the three pictures from a minimum value of 0 (black) to a maximum value of 3 (red) (A); quantitative analysis shows the ratio between the SUV in the tumour leg and in the control reference (control leg). Data represent means +/- SEM (n ≥ 3). **<0.01

Tracer	Mouse models	Sensitivity (%)	No. of mice with detectable tumors/ Total No. of mice
^{18}F-FMISO	Osteolytic (143B)	100%	15/15
^{18}F-FMISO	Osteoblastic (LM8)	86%	6/7
^{18}F-FMISO	Osteoblastic (SaOS-2/Caprin-1)	100%	5/5

Table 3.3: Sensitivity of PET imaging in detection of primary tumour with ^{18}F-FMISO

3.1.4 ^{18}F-FLUORIDE UPTAKE IN THE THREE OS MOUSE MODELS

^{18}F-Fluoride uptake of the lesions, indicating bone remodelling, differed strongly among the three OS mouse models (Fig. 3.6A). In the osteolytic 143B model ^{18}F-Fluoride uptake in primary tumours was detectable in 7 out of 15 mice (sensitivity 47%, Table 3.4). The slightly osteoblastic model, LM8, displayed a higher ^{18}F-Fluoride uptake resulting in a sensitivity of 75%. Most pronounced uptake was detected in the highly osteoblastic SaOS-2/Caprin-1 model. The analysis of the ratios between tumour and control SUVs confirmed the visual examination: SUV ratios were 2.28±0.39 in the SaOS-2/Caprin-1 model, 1.32±0.15 in the LM8 model and 1.08±0.13 in the 143B model (Table 3.1). The differences in SUV ratios were significant between SaOS-2/Caprin-1 and LM8 (p<0.05) and between SaOS-2/Caprin-1 and 143B (p<0.01) (Fig. 3.6B).

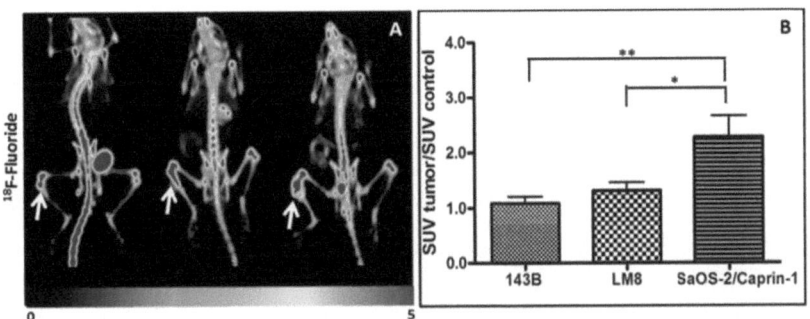

Fig.3.6: Representative PET coronal images of mice from the three different models injected with ^{18}F-Fluoride; the coloured pictures have a fixed SUV value for all the three pictures from a minimum value of 0 (black) to a maximum value of 5 (red) (A); quantitative analysis shows the ratio between the SUV in the tumour leg and in the control reference (control leg). Data represent means +/- SEM (n ≥ 3). *<0.05; **<0.01

Tracer	Mouse models	Sensitivity (%)	No. of mice with detectable tumors/ Total No. of mice
^{18}F- Fluoride	Osteolytic (143B)	47%	7/15
^{18}F- Fluoride	Osteoblastic (LM8)	87.5%	7/8
^{18}F- Fluoride	Osteoblastic (SaOS-2/Caprin-1)	100%	4/4

Table 3.4: Sensitivity of PET imaging in detection of primary tumour with ^{18}F-Fluoride

3.1.5 IMMUNOHISTOCHEMISTRY OF OS MODELS USED IN THIS STUDY

Trichromic Goldner staining and immunohistochemistry for Ki67 and CaIX were performed to validate the differences in PET tracers uptake among the three OS mouse models. Results are shown in Fig.12. Ki67 is a nuclear marker of proliferation that is expressed in cells that are undergoing the cell cycle. In our three OS mouse models Ki67 index correlates significantly with the ^{18}F-FDG uptake (Fig.3.7) and therefore we used this marker to confirm the different extent of glucose metabolism present in 143B, LM8 and SaOS-2/Caprin-1 tumour models.

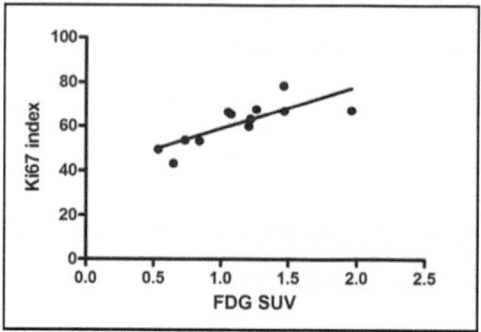

Fig.3.7: Correlation between the Ki67 index and the ^{18}F-FDG SUV in representative mice from the three OS mouse models (n=12 among which 9 mice from 143B model, 2 from LM8 model and 1 from SaOS-2/Caprin-1 model; R^2=0.62; p<0.005).

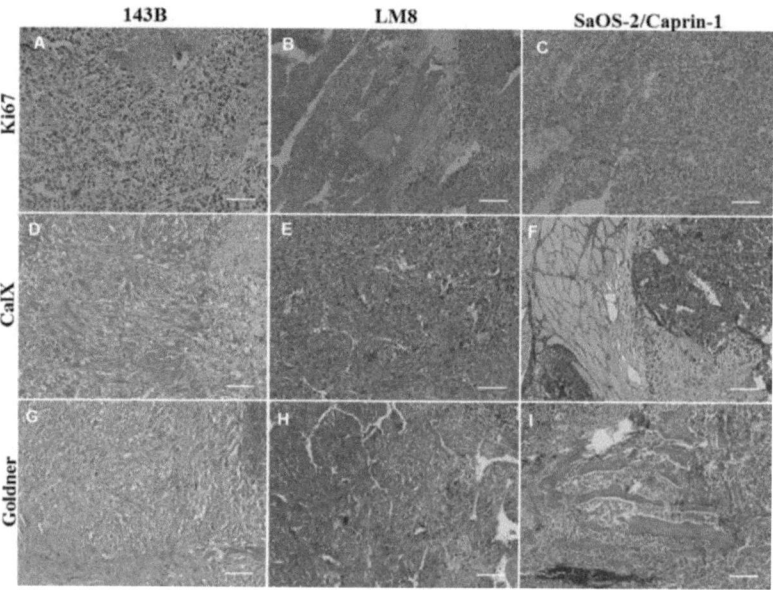

Fig.3.8: IHC (Ki67 and CaIX) and histology (Goldner staining) on the tumour sections derived from the different mouse models. Qualitatively tumour tissue staining confirms the PET imaging results: 143B model shows higher percentage of Ki67 positive cells (A)

compared to LM8 and SaOS-2/Caprin-1 models (B, C); SaOS-2/Caprin-1 model displays major extent of immature bone (F) compared to LM8 and 143B models (D, E); 143B and SaOS-2/Caprin-1 models have larger areas of CaIX positivity (G; I) compared to LM8 model (H). Scale bars: 100µm.

Indeed we could detect high presence of Ki67 positive cells in the 143B osteolytic tumour model while LM8 moderetaly osteoblastic and SaOS-2/Caprin-1 strongly osteoblastic models display a lower extent of positivity (Fig.3.8A, B and C).

CaIX is a marker of hypoxia and we used it to compare the degree of hypoxia in the three OS mouse models. The LM8 tumours showed lowest staining for CaIX (Fig.3.8D, E and F) which is in agreement with the low accumulation of ^{18}F-FMISO in this tumour model. The other two OS mouse models, 143B and SaOS-2/Caprin-1, display a much higher percentage of CaIX positive cells as well as high uptake of ^{18}F-FMISO.

As well as for Ki67, we decided to calculate the CaIX index and correlate it with ^{18}F-FMISO SUV (Fig.3.9). No correlation was found between the CaIX index and the ^{18}F-FMISO uptake which might depend on a subjective mistake in counting the positive cells, on the need of increasing the samples number or on the heterogeneity of hypoxic regions within the tumour that could be missed or overestimated in each representative slide.

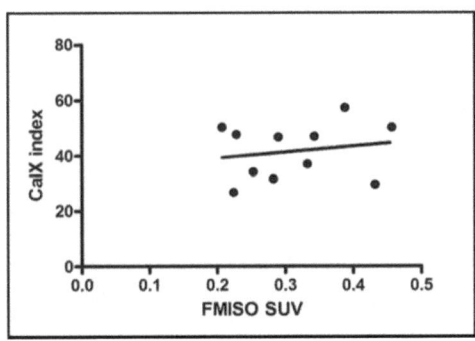

Fig.3.9: Correlation between the CaIX index and the ^{18}F-FMISO SUV in representative mice from the three OS mouse models (n=11 among which 9 mice from 143B model, 1 from LM8 model and 1 from SaOS-2/Caprin-1 model).

Another aspect that we evaluated in regard to hypoxia is the vessel density which can correlate with the degree of hypoxia present in the tumour section. Therefore we calculated the vessel area in percentage compared to the total area of the tumour sections that we analysed and we observed a tendency to a higher vessel area in the LM8 mouse model (Fig. 3.10) compared to the other two models. SaOS-2/Caprin-1 and 143B models show respectively 2.6 and 1.8 times less vessel area than LM8 model that showed among the three models also the lowest ratio in ^{18}F-FMISO uptake between tumour and control leg (Fig.3.5B).

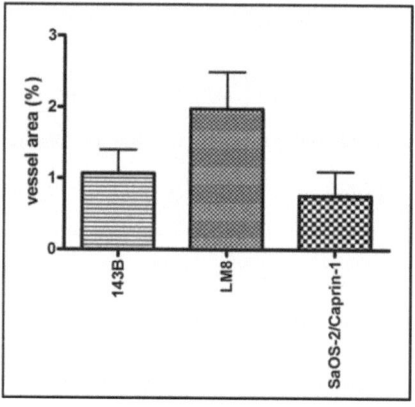

Fig.3.10: Analysis of the vessel area (%) in the three mouse models: values are represented as means ± SEM.

Trichromic Goldner staining was used to detect mineralised bone matrix that can explain the increased ^{18}F-Fluoride uptake, which is dependent on the area of mineralisation. The osteolytic 143B tumour section showed low immature bone, osteoid, and mineralised bone production from the tumour cells. The slight osteoblastic LM8 model had higher production of osteoid (red) and mineralised bone without an organised structure (blue) and SaOS-2/Caprin-1 tumours had highest presence of bone secreted by tumour cells (Fig. 3.8G, H and I). This was in good agreement with the findings from ^{18}F-Fluoride PET.

3.1.6 DETECTION OF METASTASES VIA PET IMAGING

In this study we also evaluated the possibility to detect lung metastases since the lung is the main organ of metastases in human patients. 143B and SaOS-2/Caprin-1 models develop only lung metastases with high reliability while LM8 model metastasises preferentially in kidney, ovary, liver and only to a less extent in the lung therefore we focused only on the first two models.

In the case of 143B model we could not detect any pulmonary nodules on the PET slides. Therefore we decided to perform autoradiography directly on the lung sections of one mouse using ^{18}F-FDG, ^{18}F-FMISO and ^{18}F-Fluoride after performing the whole-body PET scan. In the autoradiography it can be noted that the three tracers have different background reference. ^{18}F-FDG shows the highest background due to the physiological uptake in the lung. ^{18}F-Fluoride displays no uptake in the negative control lung which is expectable since ^{18}F-Fluoride is specific for bone remodelling. We could detect signal using the ^{18}F-FDG and ^{18}F-Fluoride that could be correlated with the presence of metastases confirmed via H&E (Fig. 3.11). Surprisingly while with ^{18}F-FDG, there was a reliable correlation between tracer signal and presence of pulmonary nodules, with ^{18}F-Fluoride, we could see that only some metastatic nodules could be detected but not all. Therefore we decided to perform the Goldner staining to verify whether the ^{18}F-Fluoride uptake might depend on the presence of mineralised matrix that is produced by the tumour cells. Indeed when we performed Goldner staining we could see that some lung nodules did not contain any mineralised matrix (Fig. 3.12A) while some others displayed a clear mineralisation in the middle of the metastases (Fig. 3.12B) which can explain why some metastatic nodules take up ^{18}F-Fluoride.

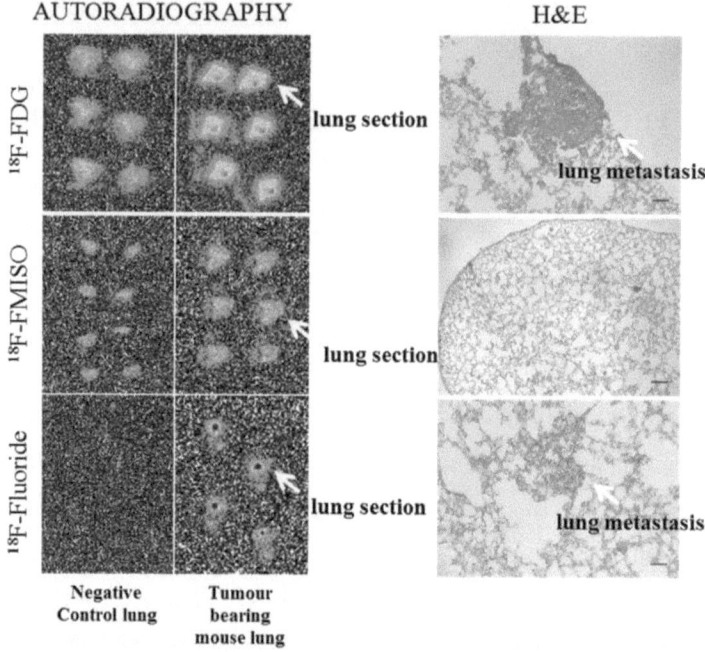

Fig.3.11: Representative autoradiography pictures from lung section of mice that did not receive tumour cells injection but that were injected with ^{18}F-FDG, ^{18}F-FMISO or ^{18}F-Fluoride (negative control) and of tumour bearing mice (tumour bearing mouse lung) (Left panel); H&E on lung sections after autoradiography of the tumour bearing mice: the presence of lung metastases is displayed with a white arrow (Right panel). Scale bars: 100um.

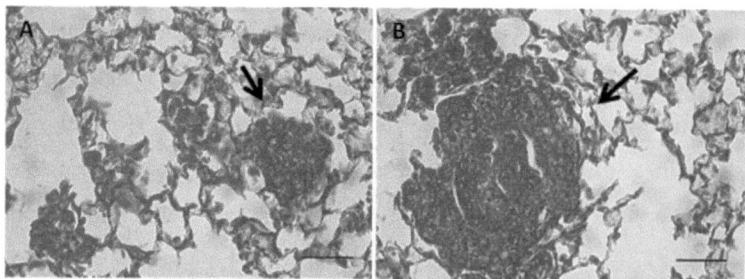

Fig.3.12: Goldner staining of lung sections from mice that received ^{18}F-Fluoride: in A no mineralised matrix is visible while in B a mineralised area is visible in the centre of the metastatic nodule. Scale bars: 50um

With ^{18}F-FMISO, no metastases could be visualised by autoradiography and by H&E.

In the SaOS-2 model it was difficult to detect metastases via PET imaging though the metastases were quite big. Moreover we could not detect any metastases by autoradiography due also to the absence of lung metastases bigger than 0.2mm in diameter in the mice used for this analysis.

3.2 DISCUSSION & OUTLOOK

Biomedical imaging is used, so far, for screening and for monitoring of disease progression, but in the future it will hopefully be applicable also for the diagnosis of developing tumours at an early stage and for the stratification of the treatment based on the imaging results of metabolic imaging. In this study, we analysed the potential of PET imaging as a tool for the characterisation of tumour different tumour types in well-defined intratibial osteosarcoma (OS) mouse models.

Nowadays PET imaging in the clinic is mainly used for monitoring treatment success in soft tissue and bone sarcomas. Indeed it was shown by Benz et al. that in patients with high-grade soft tissue tumours a cut-off of a 35% reduction of the SUV in tumours imaged with ^{18}F-FDG predicts a good response to treatment (Benz, Czernin et al. 2009). Consistent with these findings, Kim et al. showed that in patients with OS or with Ewing sarcoma ^{18}F-FDG-PET can be used for an assessment of the response to chemotherapy before surgery: they compared the ^{18}F-FDG-SUV in the tumour region in 23 patients before (SUV$_1$) and after chemotherapy (SUV$_2$) and they found that the SUV$_2$ was significantly (p<0.04) higher in the poor than in the good responders (Kim, Kim et al. 2011). Recently PET gained attention as a tool for the characterisation of metabolic processes occurring in tumours, which are more informative than anatomical changes of tumour size, visualised and monitored by CT and x-ray (Juweid and Cheson 2006).

In our study we explored the possibility of using PET imaging for the characterisation of different tumour phenotypes in well-defined OS mouse models.

In an initial study, we compared the uptake of 6 PET tracers that are available and approved for clinical use by the Food and Drug Administration (FDA). The tracers assessed were ^{18}F-FDG, ^{18}F-Fluoride, ^{18}F-FMISO, ^{18}F-FET, ^{18}F-FLT and

^{18}F-FCH in an orthotopic OS mouse model generated by intratibial injection of human OS 143B. ^{18}F-FDG was chosen because it is currently the most commonly used tracer in OS, ^{18}F-Fluoride is the most specific tracer for bone formation and remodelling and ^{18}F-FMISO was reported to indicate hypoxia in spontaneous canine osteosarcoma; the other tracers were chosen because they are indicators of distinct metabolic processes, which can affect the progression of the primary tumour to a different extent and because they have so far not been used in bone related diseases.

We found that in the intratibial osteolytic 143B cell-derived OS model ^{18}F-FCH and ^{18}F-FET uptake was only detectable in 2 out of 4 mice; ^{18}F-FLT and ^{18}F-Fluoride uptake was found to be lower in the tumour than in the control leg. However, a robust uptake of ^{18}F-FDG and of ^{18}F-FMISO was observed in the 143B OS mouse model. The remarkable intra-tracer and inter-tracer differences in tumour uptake in 143B OS indicated considerable tumour heterogeneity similar to that observed in human tumours.

A similar study has been performed by Ebenhan et al. where they compared the uptake of ^{18}F-FDG, ^{18}F-FCH, ^{18}F-FLT, ^{18}F-FET uptake in rodent lung, colon, breast cancers and in fibrosarcoma and melanoma. ^{18}F-FCH uptake was low or non-detectable in any of the models. A moderate uptake of ^{18}F-FET was observed in all models. ^{18}F-FLT exhibited the highest uptake in all the different tumours models and was considered as a generic tracer (Ebenhan, Honer et al. 2009). ^{18}F-FDG uptake was high only in the syngeneic orthotopic models (fibrosarcoma and melanoma) which were shown to be more extensively vascularised than xenograft models, suggesting that ^{18}F-FDG uptake might depend on the blood supply of the tumour.

Based on the results of our pilot study with six tracers in the intratibial osteolytic human 143B OS cell line derived model, we performed a follow-up study in three different intratibial OS mouse models with three distinct and defined OS phenotypes. The three phenotypes included the osteolytic 143B cell-derived tumours, the mildly osteoblastic tumours derived from mouse LM8 OS cells and tumours with a pronounced osteoblastic phenotype generated from SaOS-2/Caprin-1 OS cells that stably overexpressed Caprin-1, which accelerated primary tumour growth (Sabile, Arlt et al. 2013). We limited the number of tracers to the three most reliable ones in the pilot study. We used ^{18}F-FMISO, indicating hypoxia, ^{18}F-Fluoride, indicating bone formation and remodelling and the clinically so far most frequently used ^{18}F-FDG as a reference. We speculated that with the three selected tracers we would be able to visualise and distinguish the three distinct phenotypes defined by the three models.

A robust uptake of ^{18}F-Fluoride by SaOS-2/Caprin-1 cell-derived tumours, compared to the ^{18}F-Fluoride uptake by the LM8 and 143B cell-derived tumours, was in good agreement with the pronounced osteoblastic phenotype of the SaOS-2/Caprin-1 model. Our findings were also consistent with those reported by Hsu et al who investigated ^{18}F-Fluoride and ^{18}F-FDG uptake over time in osteolytic, osteoblastic and mixed type bone metastatic lesions in three distinct prostate cancer models. These osteblastic lesions showed an increase in ^{18}F-Fluoride and ^{18}F-FDG uptake over time in parallel with the growth of the bone metastases. In the osteolytic lesions, on the other hand, an increase in tracer uptake over time was only observed for ^{18}F-FDG. This is again largely consistent with our observations of a 3.8-fold higher tumour-specific uptake of ^{18}F-FDG in the osteolytic 143B OS model than in the osteoblastic LM8 and SaOS-2/Caprin-1 cell line-derived models. Thus, in both studies osteoblastic and osteolytic bone lesions could be distinguished by differential ^{18}F-Fluoride and ^{18}F-FDG uptake.

^{18}F-FMISO, the third tracer used in the three distinct mouse models, was of particular interest because it indicates hypoxic conditions in tissue and because hypoxia in tumour tissue is considered to promote chemoresistance, which has been shown to be related to a poor prognosis for OS patients in particular (Brahimi-Horn, Berra et al. 2001; Cosse and Michiels 2008). We observed remarkable uptake of ^{18}F-FMISO indicating hypoxic tumour tissue in the SaOS-2/Caprin-1 osteoblastic and in the 143B osteolytic models and less labelling in the mildly osteblastic LM8 tumours, likely due to their smaller volume than that of the primary tumours recognised in the other two models. Interestingly, we also observed a higher density of blood vessels in the LM8 than in the 143B and SaOS-2/Caprin-1 cell-derived tumours. These findings are in agreement with those of a study that demonstrated that the hypoxic regions within a tumour are usually less vascularised than the oxygenated areas. This was observed in immunodeficient mice that received human prostate cells injected into the right flank. These cells expressed EGFP only under hypoxic conditions (Raman, Artemov et al. 2006). Vascularisation was visualised with a fiber optic oxygen probe used for the measurement of the oxygen pressure and with MRI for vessel imaging. Hypoxic tumour areas contained fewer vessels, but the vessels were more permeable. Altogether, the authors concluded that the limited blood supply caused hypoxia and consequently enhanced the expression of VEGF, which increased the permeability of the vessels in these tumour areas (Raman, Artemov et al. 2006).

In order to further verify our interpretation of the PET imaging data obtained with ^{18}F-Fluoride, ^{18}F-FDG and ^{18}F-FMISO, we also performed an *ex vivo* analysis of primary tumour tissue by histology and immunohistochemistry. Tumour tissue hypoxia was assessed by staining of CaIX, which, mediated by HIF1-α, is induced by hypoxia. In a syngeneic and subcutaneous model of rhabdomyosarcoma it was shown that the expression of CaIX correlated with the uptake of ^{18}F-FMISO (Dubois, Landuyt et al. 2004; Potter and Harris 2004).

In our study we observed considerably large areas of hypoxia in the 143B and SaOS-2/Caprin-1 cell line-derived tumours, but there was no correlation between tracer uptake and the percentage of stained tumour cells per analysed area. This result might be explained by the low number of tissue samples that could be evaluated.

High glucose uptake and metabolism in tumour tissue imaged by ^{18}F-FDG PET is considered as an indicator of aggressive tumour proliferation and we indeed observed a correlation between in *vivo* ^{18}F-FDG uptake and the percentage of proliferating cells visualised *ex vivo* in tumour tissue sections by immunostaining of the proliferation marker Ki67 (Ki67 index). This result is consistent with the findings of Yamamoto et al. who reported a significant (R^2=0.81; p<0.0001) correlation between Ki67 index and FDG SUV in non-small cell lung cancer tissue of 18 patients (Yamamoto, Nishiyama et al. 2007).

In order to histologically assess the formation and remodelling of bone-like structures in osteoblastic primary tumour tissue in the here investigated OS models and to compare it with the uptake of ^{18}F-Fluoride examined by PET, we perfomed Goldner staining of mineralised and non-mineralised bone matrix in primary tumour tissue sections. The results were as expected and showed much larger areas of mineralised and non-mineralised bone matrix in osteoblastic SaOS-2/Caprin-1 and the LM8 cell line-derived tumours than in the osteolytic tumours formed by 143B cells. Thus, taken together, the histological and immunohistochemical analysis of primary OS tumour tissue confirmed the findings of the *in vivo* characterisation by PET of the different tumour types generated in the three OS mouse models investigated here.

Despite the promising results in this study, there are certain limitations. We used preclinical models that do not fully reproduce the human disease and the heterogeneity of the tumour tissue with osteolytic and osteoblastic lesions. In addition, PET imaging was performed at a late stage of tumour development when the different phenotypes were well established. Nevertheless, the here achieved successful characterisation of different tumour phenotypes in experimental OS by PET with three different tracers is encouraging. Such an approach may in the future help the clinicians to define the predominant phenotype in an OS primary tumour and to choose more tumour-type specific combinations of currently accepted and future novel therapeutic compounds.

Indeed new approaches are being investigated on the use of drugs that target glucose metabolism, hypoxia and bone remodelling processes (osteoblastic or osteolytic) (Guise, Mohammad et al. 2006; Akiyama, Dass et al. 2008; Wilson and Hay 2011; Jones and Schulze 2012). Future preclinical studies need to address the real benefit of the here proposed strategies in comparison to currently used treatment regimens with the final aim of increasing the survival rate of the OS patients.

The additional goal of visualising metastatic lesions in the present study was not reached with any of the tracers. Interestingly, there are not many studies that investigated the role of PET imaging for the detection of metastases in tumour models in mice. Recently, Deroose et al. were able to detect lung metastases 45 days after intravenous injection of melanoma cells in nude mice (Deroose, De et al. 2007). Consistent with these results, Franzius et al. reported that lung and soft tissue metastases could be detected by PET imaging with ^{18}F-FDG and osseous metastases in a xenograft mouse model of Ewing sarcoma were detectable with ^{18}F-Fluoride and ^{18}F-FDG (Franzius, Hotfilder et al. 2006). However, kidney and ovary metastases remained undetectable. The authors claimed that a high background of ^{18}F-FDG uptake in abdomen and kidney might have hindered the visualisation of metastases in those organs. In our study, we need to consider the following explanations for the non-successful visualisation of metastases in the lung that were easily recognized *ex vivo*: the resolution of our PET scanner was limited and the background of ^{18}F-FDG uptake in the lung was also high. Interestingly, we were able to detect metastatic foci in 143B model with *ex vivo* autoradiography of ^{18}F-FDG and ^{18}F-Fluoride in lung tissue sections. The observed uptake of ^{18}F-Fluoride was a surprising finding, but, upon histological analysis of the tissue, the uptake turned out to reflect the formation of mineralised matrix observed in some but not all of the metastatic lesions confirmed by Goldner staining.

The autoradiography of lung sections of mice that received ^{18}F-FDG revealed a linear correlation between the autoradiographic signals and the metastatic load assessed by histology. The finding indicated an elevated glucose metabolism in metastatic lesions compared to normal lung tissue.

The mouse subjected to ^{18}F-FMISO PET did not develop any metastases and further studies are needed to evaluate metastatic imaging with ^{18}F-FMISO.

Altogether, we conclude that the sensitivity and the resolution of the scanner available to us in this study were too low for imaging of lung metastatic lesions in mice.

These promising results of the here reported PET studies set the stage for further investigations on two aspects.

In the OS mouse models where the PET imaging successfully characterised the biology of the different OS phenotypes, new treatment approaches can be tested that, based on the PET results, target glucose metabolism, hypoxia or bone remodelling/destruction. These drugs are at the moment available in the market but they have not been tested so far in OS hence the first step would be to try some of the selected drugs that block the above mentioned processes. In order to get a reliable result, the selected drugs shall be tested only on the two xenograft mouse models used in this project since these two models closely reflect the human progression of the disease and therefore it would be easier to interpret the effect of the drug and design the further results in humans.

Examples of possible drugs that can be tested in OS mouse models can be divided in three classes: inhibitors of glucose metabolism; inhibitors of hypoxia and regulators of the imbalance between the osteoclastic and osteoblastic activity. To the first class 2-deoxyglucose (2DG) and dichloroacetate (DCA) belong: the first one enters the cells via the glucose transporters, gets phosphorylated by the hexokinase and saturates this enzyme (Jones and Schulze 2012); the second one reactivates the tricarboxylic acid cycle (TCA cycle) whose activity is reduced in many cancers. DCA has shown to restore the passage of pyruvate in the mitochondria and consequently the TCA cycle inducing apoptosis and tumour shrinkage (Jones and Schulze 2012).

Concerning the second class of drugs, many classes of small drugs have been developed to target specifically hypoxic tumour cells but not all produced promising results. One class of compounds, that is already in multiple phase I and II trials with favourable results, is represented by bioreductive prodrugs which are usually undergoing a double reduction which is reversible in oxic cells but not in hypoxic cells where these prodrugs will generate toxic radicals. Example of prodrug is N-oxyde tirapazamine (TPZ) which is 50-200 folds more toxic to hypoxic than oxic cells in culture and TH-302 which is a 2-nitroimidazole-based nitrogen mustard prodrug (Wilson and Hay 2011).

Finally concerning drugs which might regulate the imbalance between the osteoclastic and osteoblastic activity, promising findings have been published. In the case of osteolytic lesions, bisphosphonates have been shown to inhibit the bone resorption and therefore they are already used in the treatment and prevention of osteoporosis, Paget's disease, tumour metastases in bone and skeletal disorders. The zolendronic acid (ZOL), a third generation bisphosphonates, displayed also strong inhibitory effect on the osteolysis and on the formation of lung metastases (Akiyama, Dass et al. 2008).

In the case of drugs which inhibit bone formation an example is Atrasentan, which is an antagonist of ET-1 receptor (ET_AR), reduced the incidence of bone metastases though it is not yet clear how ET-1 is involved in osteoblasts stimulation (Guise, Mohammad et al. 2006).

Once the pre-clinical studies will give promising results with these new drugs, the next step is to investigate the role of PET imaging in humans. There are many parameters that need to be verified: first of all the feasibility of the use of three tracers in human patients in different days; secondly the possibility of combining PET with the histology and CT for the final planning of tumour type specific treatment modalities. This can only be achieved if the new drugs under investigation, which specifically target cellular processes, are found to be effective in experimental OS.

3.3 MATERIAL & METHODS

3.3.1 CELL LINES

The human OS cell line 143B was obtained from the European Collection of Cell Cultures (Salisbury, UK); murine LM8 OS cells were a kind gift from T. Ueda (Osaka National Hospital, Osaka, Japan). These cell lines were transduced with retroviral particles producing *LacZ* (including the neomycin resistance) as described by Arlt et al. and all the cells were kept in culture with DMEM-HamF12 (PAA GmbH, Cölbe, Germany) supplemented with 10% fetal calf serum and 1.2 mgkg^{-1} of Neomycin (Invitrogen, Zug, Switzerland) for the 143B cells and 2 mgkg^{-1} of Neomycin for the SaOS-2 cells (Arlt, Banke et al. 2011). Human SaOS-2 cells transduced with a Caprin producing vector (including neomycin resistance) were kindly provided by Dr. Sabile (Sabile et al, Orthopedic University Clinic Balgrist Zurich, Switzerland-submitted paper) and named SaOS-2/Caprin-1. These transformed SaOS-2 cells were cultured in the same medium described above supplemented with 0.6mgmL^{-1} Neomycin.

3.3.2 MOUSE MODELS

Animal care and experiments were conducted in accordance with Swiss Animal Welfare legislation and were approved by the Veterinary Office of Canton Zurich, Zurich, Switzerland. Six to eight weeks old female immunosuppressed SCID and immunocompetent C3H mice were purchased from Charles River Laboratories (Sulzfeld, Germany), kept in pathogen-free conditions. The mice have always reached the mouse facility at least 1 week before starting any experiment. Before injection, cells were detached with trypsin-ethylenediaminetetraacetic acid (EDTA), washed twice with PBS containing 0.05% EDTA and finally re-suspended in PBS-EDTA at a final concentration of 5x10^7 cellsml^{-1}. An aliquot of the cell suspension (10 µl) was orthotopically injected into the medullar cavity of the left tibia of the micePrimary tumour development was monitored by X-ray (Faxitron, Arizona, USA) weekly for 143B and LM8 cell-derived tumours and once every two weeks for SaOS-2/Caprin-1 cell derived tumours. The 143B injected mice were sacrificed between days 21 and 23 after tumour cell inoculation; LM8 mice on day 27 and SaOS-2/Caprin-1 mice between days 61 and 115. As said in the paragraph "OS animal models" the SaOS-2/Caprin-1 tumour model showed a high heterogeneity in tumour development.

3.3.3 PRIMARY TUMOUR VOLUME

Primary tumour was measured with a caliper once per week in the orthotopic 143B and subcutaneous and orthotopic LM8 models and twice per week in the SaOS-2/Caprin-1 model. The volumes of the tumour and the control legs were calculated with the following formula: $(LxW^2)/2$ where L is the length and W is the width of the leg. The control leg is used as a reference of the physiological bone growth therefore the final tumour volume is calculates as follows: $V_{tum} - V_{ctr}$ where V_{tum} is the volume in the tumour leg and V_{ctr} is the volume in the control leg.

3.3.4 RADIOTRACER SYNTHESIS

The radiotracer synthesis was performed at ETH Hönggerberg in the Animal Imaging Center by Cindy Fischer and Mathias Nobst.
^{18}F-FDG was obtained from routine productions for clinical use at the University Hospital Zurich.

3.3.5 PET SCANS

Animals were transported to the PET facility for acclimatisation 5-7 days before the first PET scan. PET tracers were injected into a lateral tail vein in mice, except for ^{18}F-FDG scans in 143B and SaOS-2/Caprin-1 mice (see below). ^{18}F-Fluoride (20.4-46.7 MBq) and ^{18}F-FMISO (16.1-44.6 MBq) were injected 90 min before the scan started; ^{18}F-FDG (9.5-16.4 MBq) and ^{18}F-FLT (4.3-12.93MBq) 30 min before the scan started; ^{18}F-FCH (9.25-14.17 MBq) and ^{18}F-FET (7.84-13.25MBq) 15 min before the scan started. Ten minutes before scan started, mice were anaesthetised in an incubation chamber containing 5% isoflurane (Abbott, North Chicago, USA) in a 1/1 mixture of air and oxygen and transferred to the PET/CT scanner. To reduce muscle uptake of ^{18}F-FDG in mice, anaesthesia with isoflurane was initiated as described above on a heating pad (Harvard Apparatus, Massachusetts, USA) 10 min before ^{18}F-FDG injection. Mice were then kept at 37°C body temperature under anaesthesia until the end of the PET scan.

PET whole body images were acquired on a Vista eXplore preclinical PET/CT scanner (GE Healthcare, Glattbrugg, Switzerland) with 2 bed positions, 15 min each. During the scans, mice were kept in anaesthesia with 2-3% isuoflurane in air/oxygen (200cc/min each) and monitored as previously described (Honer, Bruhlmeier et al. 2004; Honer, Hengerer et al. 2006). PET was followed by a Computed Tomography (CT) scan in some of the experiments.

The reconstruction and the coronal images were performed as described in Mumprecht et al. (Mumprecht, Honer et al. 2010). Coronal images in false colours were obtained with the PMOD software (PMOD technologies Ltd., Zurich, Switzerland). For direct comparison, the colour scales were normalised to injected radioactivity dose and body weight of the animal to reveal the standardised uptake value (SUV). All SUV scales were set to maximal values of 3 for ^{18}F-FDG, 0.75 for ^{18}F-FMISO and 5 for ^{18}F-Fluoride. Scale minimum was 0 for all tracers. PET imaging with ^{18}F-FCH, ^{18}F-FET and ^{18}F-FLT was performed only in the orthotopic 143B and subcutaneous LM8 models: we did not use these tracers in the other models because the results within the 143B model were heterogeneous therefore for these tracers maximal intensity projection picture will be shown for the 143B model. For the subcutaneous model a short summary of the results will be given since we did not further investigate on this model. Mice were sacrificed after the last PET scan.

3.3.6 X-Gal STAINING

The procedure for the sacrifice of the mice is specifically optimised for the X-Gal staining that we perform on the lungs to visualise the metastases and on the legs to confirm presence of primary tumour. After intraperitoneal administration of a sleeping solution that contains 10% Ketamin, 2% Xylazin and 10% Acepromazin, mice are open at the level of abdomen and thorax in order to remove the skin and reach the heart. After cutting the vena cava 10-15ml of PBS were injected into the right ventricle of the heart in order to get rid of the blood in the lung. Afterwards a clamp is applied on the vessels below the lung and above the liver in order to fix the lung with 3% PFA: one part of the 3% PFA (1.5ml) is given via the right ventricle while 2 ml is injected through the trachea. Thereafter, another clamp is applied on the trachea. While the lung is being fixed, the primary tumour and the control leg are isolated and finally the lung is washed with PBS to clean it from the 3% PFA, excised, left in 4% PFA and stained with 5-bromo-4-chloro-3-indolyl-b-D-galactoside (X-gal) for at least 3 hrs as described previously (Arlt, Kopitz et al. 2002). Afterwards the organs are washed with PBS and kept at 4° C in 4% PFA for long-term storage. When the autoradiography of the lungs was performed we used the same procedure for the sacrifice of the mice but we used the right middle and cranial lobes for the X-Gal Staining and the right caudal and the left lobes for the autoradiography (see next paragraph).

3.3.7 QUANTITATIVE ANALYSIS AND CALCULATION OF THE SENSITIVITY

PET images were analysed drawing a region of interest (ROI) around the tumour and mirroring the area to the healthy control leg for all the PET slices where the tumour was visible: the sum of all the ROIs gives a volume of interest (VOI) and the activity concentration (counts/sec) that is a measure of the tracer accumulation in that specific VOI (Mikolajczyk, Szabatin et al. 1998). ^{18}F-FDG uptake in the control leg of the mice, that were awake during and after tracer injection, was higher than in the tumour leg. In these mice, the hip was used as control. Fused PET/CT images were used to draw the ROIs in these mice (Fig.3.13). The average radioactivity concentrations in the tumour and healthy leg/hip were expressed as SUV and their ratios were calculated and compared in the three OS mouse models. Mice with too high radioactivity in the urinary bladder were excluded from the analysis. Sensitivity was calculated as number of mice with a detectable tumour in the PET imaging slices, as concluded by visual inspection by two independent blinded researchers, divided by the total number of mice.

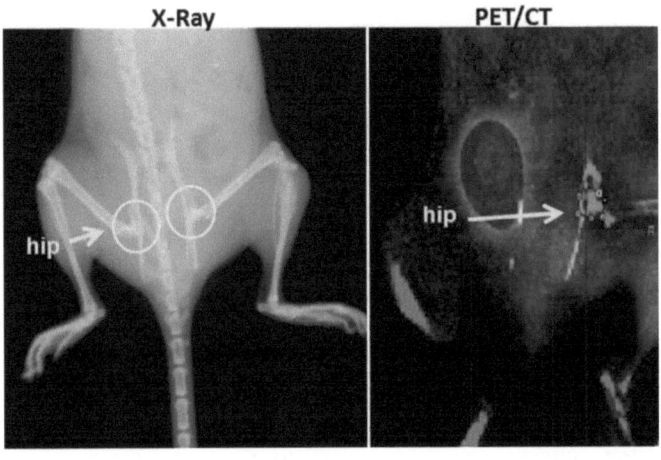

Fig.3.13: Imaging of the hip region that was used for the quantification of ^{18}F-FDG uptake in the LM8 mouse model using x-ray (left panel); example of a PET/CT fused image that was used to draw the background region of interest.

3.3.8 HISTOLOGY AND IMMUNOHISTOCHEMISTRY

Primary tumour legs, after being at 4° C with 4% PFA for at least one night, were decalcified in Osteosoft® (Merck Chemicals, Darmstadt, Germany) for 1 week, embedded in paraffin for histological and immunohistochemical stainings. In the case of the ^{18}F-FDG, ^{18}F-FMISO and ^{18}F-Fluoride where we compare the tracers' uptake in the different OS mouse models, we confirmed the results with immunohistochemistry (IHC) and histology. Therefore we used Ki67, marker of proliferation, CaIX, marker of hypoxia, and Trichrome Goldner staining were used. 6µm tumour sections were cut: histological analysis was performed with a Trichrome goldner staining kit following the instructions of the provider (Carl Roth GmbH, Karlsruhe, Germany); immunohistochemical analysis was carried out using rabbit polyclonal Ki67 Antibody (Abcam, Cambridge, UK) and rabbit polyclonal CaIX Antibody (Novus Biologicals Ltd., Cambridge, UK). The staining was visible by peroxidase-based reaction using Vectastain® Elite ABC and a substrate-chromogen system (Dako, Baar, Switzerland). All the pictures were taken via visible light and 10x lens with a Zeiss Observer.Z1 inverted microscope (Zeiss 43) (Carl Zeiss MicroImaging GmbH, Göttingen, Germany).

3.3.9 Ki67 AND CaIX INDEX

Ki67 index was calculated as the total number of Ki67 positive nuclei (Ki67 localises in the nuclei) divided by the total number of nuclei. For this analysis immunofluorescence was performed, the primary Ki67 Antibody used was the same as described in the previous paragraph while the secondary was a fluorescent rabbit antibody (Alexa-Fluor 546, Invitrogen, Zug, Switzerland). DAPI was used as our reference for counting the total number of nuclei. The total and Ki67 positive cells are counted with Image J software that analyses the number of particles. We selected three slides per mouse taken at different depth and we took seven pictures per slide using a Nikon 40x lenses equipped with an appropriate filter block specific for DAPI and for the Alexa 536 Antibody (Nikon Corporation Eclipse E600, Tokio, Japan).

CaIX index was calculated with the same formula as Ki67 index and in this case IHC was performed as described before. The analysis was performed visually counting the total number of cells with haematoxylin. In this case we selected three slides per mouse at different depth and we took six pictures per slide using the same microscope Nikon microscope mentioned above.

3.3.10 VESSEL AREA

To calculate the vessel area in each tumour model, immunofluorescence using CD31 antibody (Abcam, Cambridge, UK), marker for endothelial cells, was performed on 8um primary tumour sections: the procedure was the same as described in the previous paragraph.

For each mouse three slides from different depth were cut, stained with CD31 and three pictures per slide were taken via visible light and 10x lens with a Zeiss Observer.Z1 inverted microscope (Zeiss 43) (Carl Zeiss MicroImaging GmbH, Göttingen, Germany).

For the calculation of the vessel area we used the software Image J with a MACRO designed by Csaba Balazs (Light Microscope Center, ETH-Hönggerberg, Zurich).

The percentages of vessels present in the whole area for the three sections were averaged and for each model the analysed mice were averaged: 5 mice in the case of 143B, 2 mice in the case of LM8 and 4 mice in the case of SaOS-2/Caprin-1 model.

3.3.11 AUTORADIOGRAPHY OF THE LUNGS

To perform autoradiography on the lungs of the mice, the mice were first scanned by PET imaging and later sacrificed to perfuse and excise the organs, lungs were embedded in OCT (Optimal Cutting Temperature) embedding medium and snap-frozen. The autoradiography was performed only on lungs of mice injected with ^{18}F-FDG, ^{18}F-FMISO and ^{18}F-Fluoride: one mouse for the 143B model and one mouse for the SaOS-2/Caprin-1 model were used for each tracer. Mice lungs from LM8 model were not analysed since this model does not metastasise in the lung. Lung sections of 20 um thickness were cut and exposed to a FUJIFILM BAS-MS2025 based imaging plate and developed with a BAS 5000 phosphor imager to reveal the tracer uptake. A mouse that did not receive tumour cell injection and where the different tracers were injected was used as our background reference for the 143B and SaOS-2/Caprin-1 models. After lung autoradiography, the same sections were stained for H&E to confirm the presence of metastases.

3.3.12 STATISTICS

Statistical analysis was performed with GraphPad Prism® 5.01 Network. The ratios between the tumour and the control SUVs were compared among the three OS models by two tailed Student's *t*-test and the p-value<0.05 was evaluated as statistical significant and p-value <0.01 as highly statistical significant. The SUV values in the table were reported as Mean ± SEM.

CHAPTER 4

EVALUATION OF EFFICACY OF PHOTODYNAMIC THERAPY IN OS

Contributions
Study design: Carmen Campanile, Kerstin Reidy, Prof. Bruno Fuchs, Prof. Walter Born and Prof. Roman Muff
Experimental design: Carmen Campanile, Kerstin Reidy and Dr. Matthias JE Arlt
Experiment conduct: Carmen Campanile and Kerstin Reidy
Data analysis: Carmen Campanile
Animal project leader: Dr. Knut Husmann
Animal work: Carmen Campanile, Dr. Matthias JE Arlt, Dr. Patrick Brennecke and Dr. Ana Gvozdenovic
Technical support: Josefine Bertz, Dr. Sander Botter, Christopher Bühler and Dr. Ram Kumar
Material provider: Dr. Adam A Sabile

4.1 RESULTS

4.1.1 HIGHER PS UPTAKE IN THE HIGHLY METASTATIC OS CELL

Formulation 1 uptake was evaluated in two different OS cell systems consisting of parental low metastatic cell lines (HOS and MG63) and derivative high metastatic cell lines (143B and MG63-M8). The uptake was analysed over 24hrs and each parental cell line was compared to its derivative: y-values were calculated considering the value at 24hrs in the parental cell lines as 100%.

In both cell line systems there is a higher uptake in the highly metastatic cell line compared to the low metastatic ones which is visible already 2 hrs after PS addition (Fig. 4.1).

In order to confirm this finding we decided to focus on one cell line system (HOS-143B). For the 143B cell line system, the mouse model is also well established. In this cell line system we measured the Formulation 1 uptake in a time- and dose-dependent way. Furthermore the results were normalised to the cell area which was approximated to the one of a sphere and to the total number of cells (Fig. 4.2). In accordance with the previous findings the highly metastatic 143B cells take up more than the low metastatic cell line in a dose- and time-dependent way. Indeed an increase of intracellular concentration of Formulation 1 between 340 and 520 times could be detected 5hrs after incubation with the concentration of 10 μgml^{-1} Formulation 1. In view of these observations and of a future *in vivo* validation of PDT for the treatment of metastasising OS, additional experiments *in vitro* have been performed using only the highly metastatic 143B cell line, because, *in vivo*, different from their parental low metastatic HOS cells, these cells generate an aggressive form of OS in an intratibial xenograft model in SCID mice: metastases as in the human situation are found mainly in the lung. For the *in vivo* experiment we made use of *LacZ* tagged cells which enables us to confirm the presence of primary tumour and to visualise the metastases *ex vivo*.

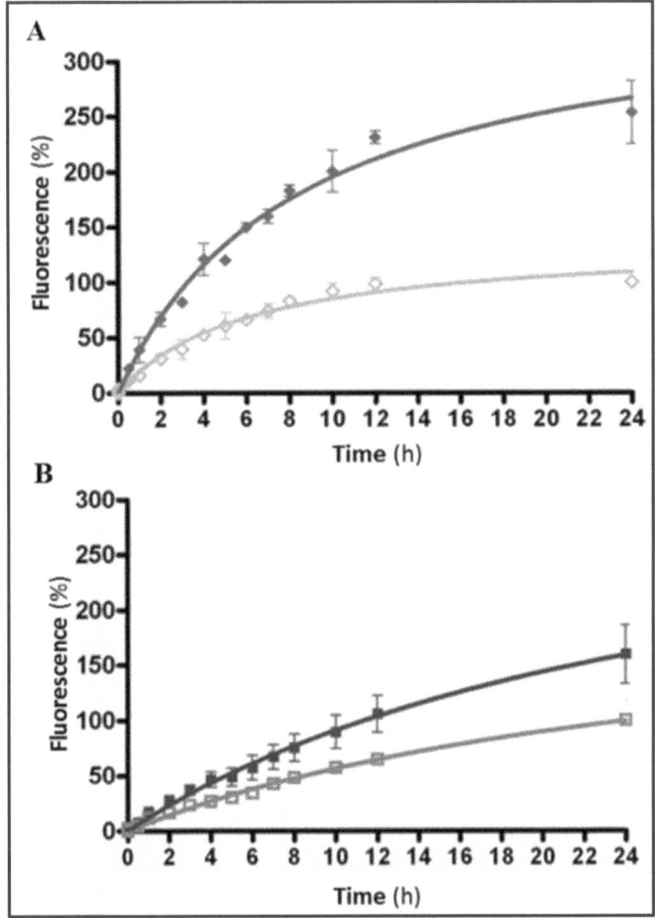

Fig.4.1: Formulation 1 uptake over time in two human OS cell line systems: (A) Parental cell line HOS (yellow) and its corresponding highly metastatic subline 143B (red), (B) Parental cell line MG-63 (light blue) and its corresponding highly metastatic subline M8 (dark blue). Data represent means +/- SEM (n ≥ 3).

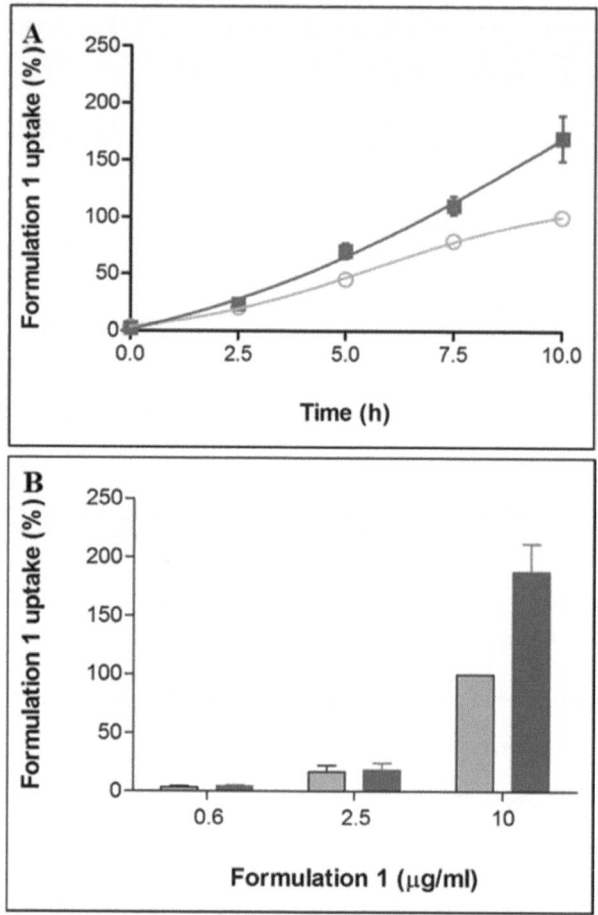

Fig.4.2: Formulation 1 uptake in one human OS cell line system: (A) Time-dependent uptake in parental cell line HOS (orange circles) and its corresponding highly metastatic subline 143B (dark red squares), (B) Dose-dependent uptake in parental cell line HOS (orange columns) and its corresponding highly metastatic subline M8 (dark red columns). Data represent means +/- SEM ($n \geq 3$).

In the 143B cells we investigated the uptake using a confocal laser scanning microscope for an evaluation of the subcellular localisation of Formulation 1, therefore the cells were incubated with increasing concentrations of Formulation 1 for 5hrs and we found that Formulation 1 mainly localises in the cytoplasm around the nuclei (Fig.4.3). The Formulation 1 uptake became detectable in all cells at a concentration of Formulation 1 ≥ 0.6 µgml^{-1}. The absence of nuclear localisation implies that the Formulation 1 is not genotoxic and therefore it does not induce further replication of newly mutated cells that can be more aggressive or resistant to further treatments.

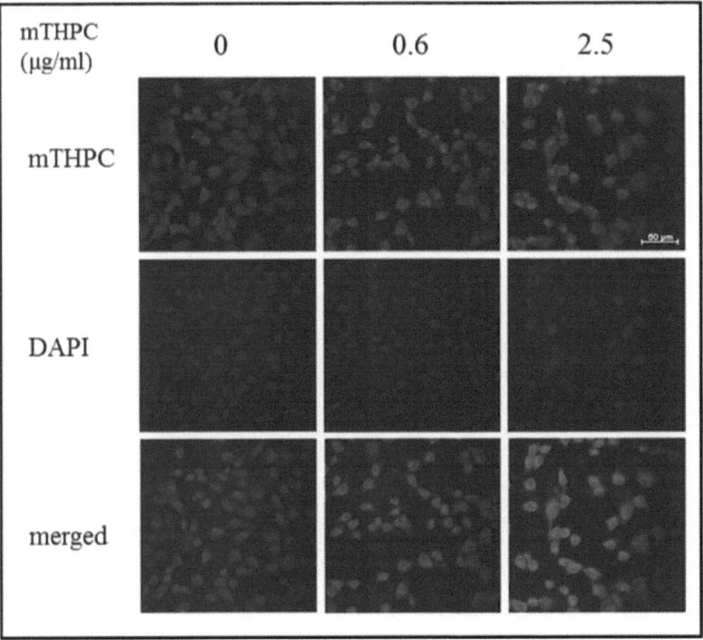

Fig.4.3: Confocal microscopy pictures of Formulation 1 uptake in 143B cells after incubation with different concentration of the PS: DAPI was used to highlight the nuclei and the merge between the two channels was used to confirm the cytosolic localisazion of Formulation 1.

4.1.2 143B CELLS ARE HIGHLY SENSITIVE TO THE PDT TREATMENT

After measuring the uptake of Formulation 1 in the 143B, we, afterwards, investigated the effect of dark- and photo-toxicity of the PS on the cells. 143B cells were incubated with different concentration of PS for 5hrs and kept either in the dark or illuminated with 652nm light. After 24hrs, a WST-1 assay was performed to measure the cell viability and cell number was counted in the different conditions (with or without illumination). Dark-toxicity in the cells was visible in a decrease of cell viability and in the number of cells only with concentrations of Formulation 1 higher than 2.5 μgml^{-1} (Fig. 4.4A). Phototoxicity was already detectable at doses of 0.075μgml^{-1} Formulation 1 or higher, independent of the light dose and was dependent on the concentration of Formulation 1 and on the laser light dose which was used (2.5, 5, 10 Jcm^{-2}) (Fig. 4.4B). Half-maximal lethal doses (LD_{50}) of Formulation 1 in cells illuminated with the different laser light dose ranged from 0.012μgml^{-1} (+/-0.008 SEM) to 0.047μgml^{-1} (+/-0.007 SEM).

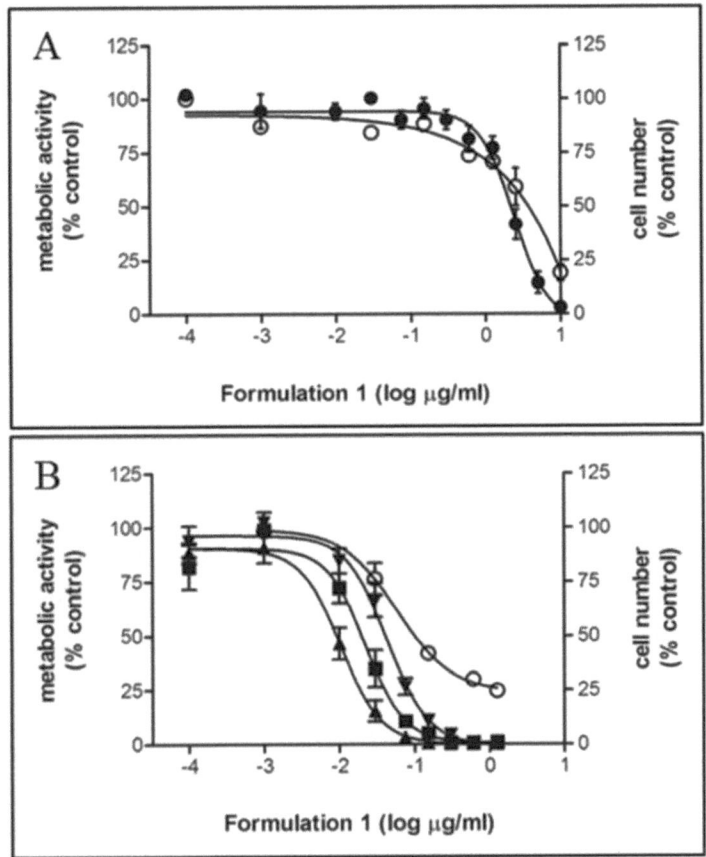

Fig.4.4: Dark- and photo-toxicity in 143B cells after incubation with different concentrations of Formulation 1 for 5hrs: (A) 143cells received PS but were kept in the dark (dark-toxicity); (B) 143B cells received PS and were illuminated with different energy dose 2.5 J/cm^2 (triangle), 5 J/cm^2 (square), 10 J/cm^2 ((photo-toxicitiy) Data represent means +/- SEM ($n \geq 3$).

In summary 143B cells are sensitive to PDT but mechanism for the induction of the cellular death is not yet clear.

4.1.3 APOPTOSIS -INDUCED PHOTO-TOXICITY

To clarify the mechanism that is responsible for the PDT-mediated cellular death, we focused on the apoptosis which is known in the literature to be the most common mechanism activated after PDT.

We checked for apoptosis activation by western blotting after incubating the cells with Formulation 1 for 5 hrs and illuminating them with 5 Jcm^{-2} energy dose. Caspases and PARP cleavage, implying apoptosis activation, occurs already 90 min after illumination with 0.6 μmml^{-1} demonstrating the direct induction of apoptosis after PDT. Indeed the cells kept in the dark did not show any unspecific activation of the apoptosis (Fig. 3.12A). To further confirm the specificity of the PDT-mediated apoptotic stimulation, we used the pan-caspase inhibitor, Z-VAD-FMK, in different concentrations (50 μM and 100 μM). The incubation with Z-VAD-FMK for 1 hr before the illumination suppressed the caspase-dependent PARP cleavage in a dose-dependent manner (Fig. 4.5).

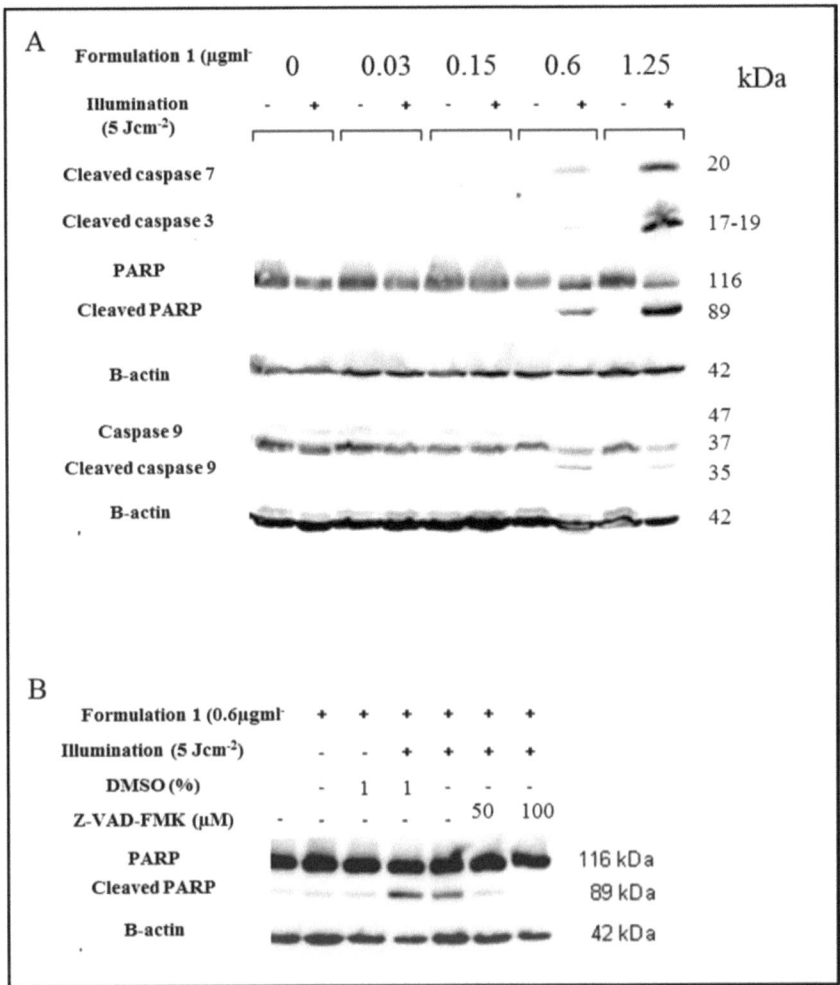

Fig.4.5: PDT-mediated apoptosis in 143B cells after 5 hrs incubation with Formulation 1: (A) cleavage of PARP, caspase 3, 7 and 9 was evaluated after 5 hrs incubation of the cells with different concentration of Formulation 1 and then illuminated with 5 Jcm^{-2} or kept in the dark; (B) inhibition of PARP PDT-induced cleavage after 1hr incubation with a pan-caspase inhibitor (Z-VAD-FMK) before illumination was performed.

4.1.4 HIGHLY SELECTIVITY OF PS UPTAKE *IN VIVO*

For the *in vivo* experiments we used Formulation 2, as mentioned in the Material & Methods part, which differs from Formulation 1 only for the dissolving liquid: 5% glucose in the case of Formulation 2 and water in the case of Formulation 1. Further we made use also of *LacZ* tagged cells for the indentification *ex vivo* of metastatic foci. The uptake *in vivo* was evaluated with two OS mouse models: one xenograft and one syngeneic where human 143B*LacZ* and murine LM8*LacZ* cells respectively were injected into the left tibia. The reason why we used both models is dependent on the *in vivo* results about the treatment studies that will be described in the next paragraph. Briefly the absence of a curative effect after PDT in the xenograft model can be related to the lack of T-cell response in the xenograft model therefore we decided to further investigate the Formulation 2 uptake in a syngeneic OS mouse model where the immune system is functional.

Concerning the 143B*LacZ* model the uptake experiment was started on day 21 when the mice showed an average volume of 106 mm^3. In general the three mice showed a similar pattern of tumour growth over the three weeks (Fig.4.6).

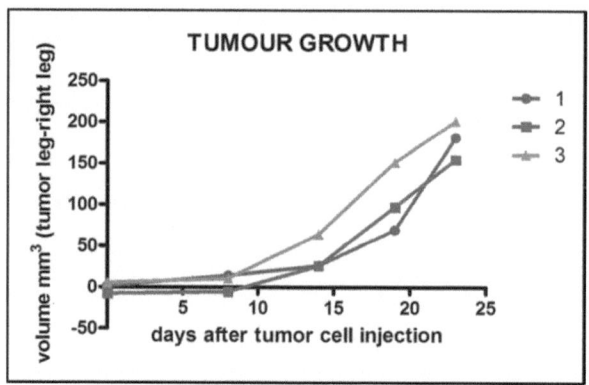

Fig.4.6: Tumour growth curve of three 143B*LacZ* tumour bearing mice.

This preliminary experiment showed high specificity of the Formulation 2 uptake in the tumour leg as it is visible in the spectrum of a representative mouse that covers wavelength from 550 to 800 nm. A single peak is visible at 652 nm with a value of 153 relative fluorescence units (RFU) (excitation wavelength of the Formulation 2) which reaches the highest value 6 hrs after intravenous injection of the PS (Fig. 4.7A).

When we combine the results of the three mice we observe that the peak of uptake at 6 hrs is consistent in all three mice. The limited number of mice used in this experiment does not allow us to evaluate and explain the second peak at 24 hrs that was also observed at biolitec research in other animal models but in our case it can also depend on the high standard deviation that we can see at 12 and 24 hrs (Fig. 4.7B).

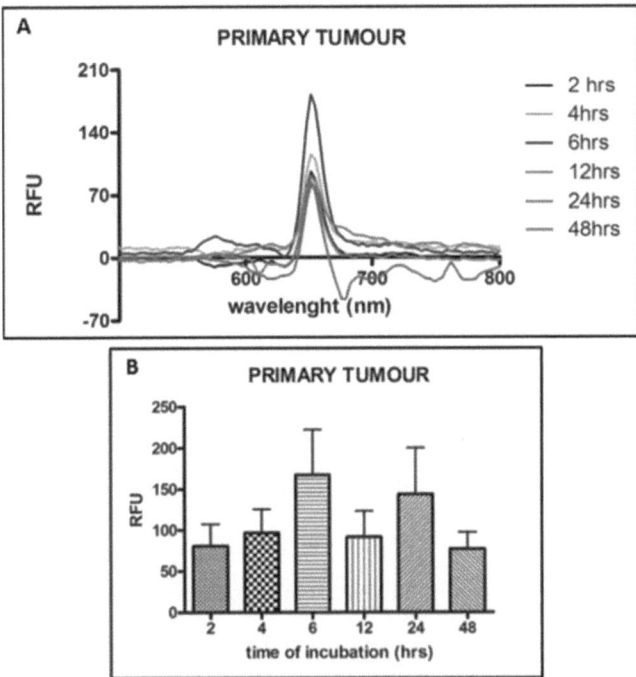

Fig.4.7: Formulation 2 uptake *in vivo* in the primary tumour leg measured at different time points after intravenous injection of 1.5 mgkg^{-1} of Foslip®: (A) spectrum from 550 to 800 nm of the tumour leg uptake of one representative mouse with the peak at 652 nm; (B) averaged relative fluorescence units (RFU) of the three mice tumour leg over time after Formulation 2 injection. Values are represented as means +/- SEM (n ≥ 3).

Formulation 2 uptake in these three mice was measured also in the liver and in the lung: both organs showed a similar pattern of uptake with the highest peak 6 hrs after Formulation 2 injection. Then slowly the RFU over time decreased. Interestingly the RFU values in the lung and liver were much lower than in the primary tumour (167.1±55.1) which underlines the high specificity of this compound for tumour tissue. Furthermore we observed also that the lung uptake was higher at 6hrs more (50.94±1.41) than the one measured in the liver (33.7±3.4) which is explainable with the presence of pulmonary metastases that can take up more Formulation 2 compared to the liver which is the organ of metabolism (Fig. 4.8).

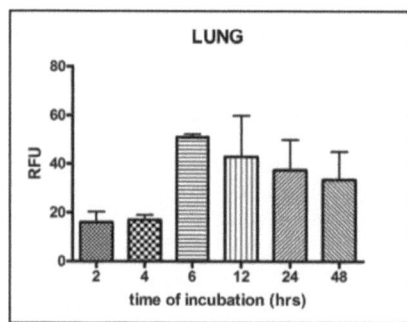 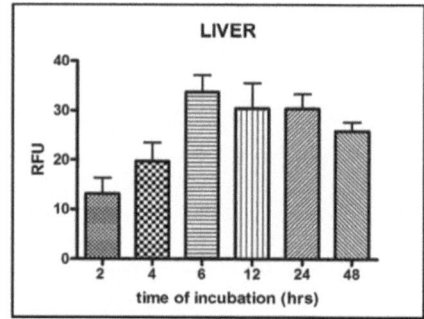

Fig.4.8: Formulation 2 uptake *in vivo* in the lung and in the liver measured at different time points after intravenous injection of 1.5 mgkg^{-1} of Formulation 2. Values are represented as means +/- SEM (n ≥ 3).

In the LM8*LacZ* model the procedure of the experiment was the same and we performed two experiments. In the first experiment the mice received the injection with 1.5 mgkg^{-1} of Formulation 2 at day 25 after tumour cell injection. The time point was chosen on the basis of the tumour growth. The mouse, named "Mouse 1", was the only one with a large tumour growth (762 mm^3) at day 20 after tumour cell injection while the "Mouse 2" showed a final tumour volume of 43.21 mm^3 (Fig. 4.9).

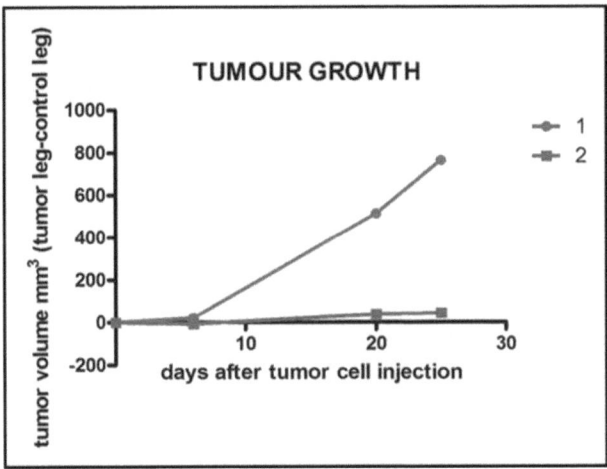

Fig.4.9: Tumour growth curve of two LM8*LacZ* tumour bearing mice.

The other 7 mice had to be sacrificed before the uptake experiment could be started due to high frequency of metastases in several organs. Despite the remarkable difference in the tumour volume the pattern of uptake in these two mice was similar in respect to each other but quite different compared to the 143B*LacZ* model because the uptake in the tumour leg increased over time till 48hrs and the RFU values are 3 to 5 fold lower compared to 143B*LacZ*. Mouse 1 shows a peak of 57.9 at 48 hrs and the mouse 2 shows a peak of 32.3 at 48 hrs (Fig. 4.10).

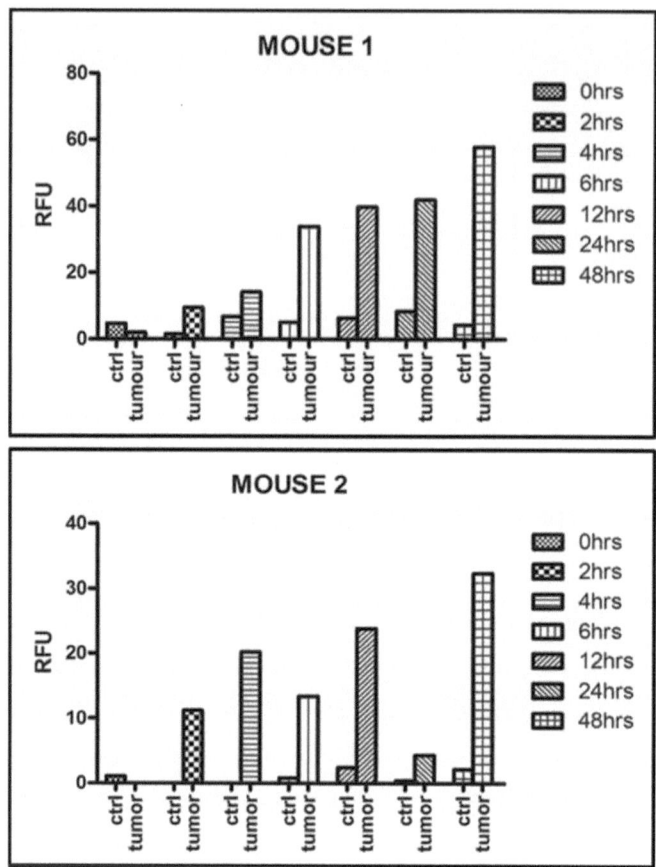

Fig.4.10: Foslip® uptake measured in the primary tumour (tumour) and healthy legs (ctrl) in two different LM8 mice (MOUSE 1 and 2).

Furthermore mouse 2 shows a drop in the value at 24 hrs but this can be due to an error in the measurement. Differently from the previous uptake experiments performed with 143B*LacZ* model, we measured here also the Formulation 2 uptake in the control healthy leg to confirm the specificity of tumour uptake and indeed we could not detect a real specific signal in any of the measured time points.

In the second experiment the uptake study was started a bit earlier, at day 21 after tumour cell injection in order to have a higher number of mice and in total we could measure 7 mice.

The tumour growth curve shows that the tumour volume of the 7 mice was quite small with an average of 21.7 mm^3 (Fig. 4.11).

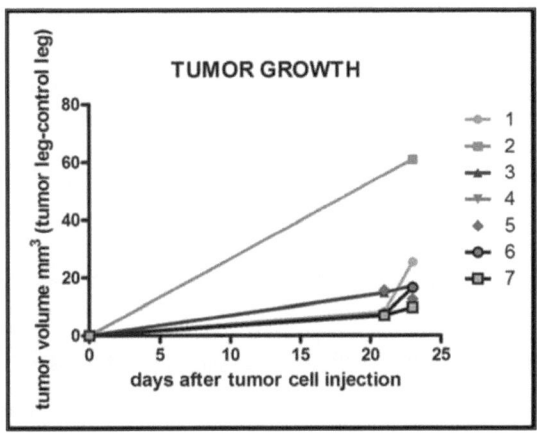

Fig.4.11: Tumour growth curve of seven LM8*LacZ* tumour bearing mice: second round of uptake experiment.

The uptake results from the second round of mice were quite different compared to the first one. The tumour leg and the control leg do not show any difference in Formulation 2 uptake and the RFU values are so low that one could consider this signal as background (Fig. 4.12).

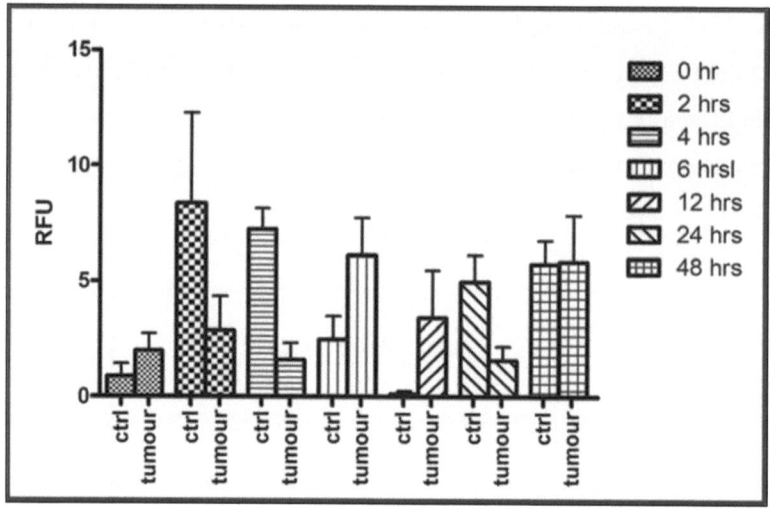

Fig.4.12: Formulation 2 uptake over 48 hrs measured in the primary tumour (tumour) and healthy legs (ctrl): average of the uptake from 7 mice is shown in the graph.

One explanation for this is that the tumour volumes of the second round of mice were so small that it was difficult to penetrate deeply with the optic fiber since the excitation wavelength is 405 nm. When the tumours are larger, there are higher chances to touch with the optic fiber the tumour region.

4.1.5 OPTIMISAZION OF PDT PROTOCOLS FOR OS TREATMENT IN MICE

Treatment studies were performed in a xenograft OS mouse model where 143B*LacZ* cells were injected in the left tibia of the mice. As mentioned in the Material & Methods (Treatment protocol) we decided to be as close as possible to the human situation starting the treatment only when the tumour could be visualised by x-ray. In the 143B*LacZ* model which is well established, we usually manage to see some lytic lesions which are typical of this model at day 14.

In the first study we selected Formulation 2 concentrations of 0.15 mgkg^{-1} which is applied successfully in various subcutaneous mouse models (see the Discussion). The main question of this first experiment was the determination of the most effective energy dose between 10 and 40 Jcm^{-2}. The study consisted of 6 groups of mice among which 4 were negative controls. The scheme is shown below (Table 4.1): the first two groups included mice that received 0.15 mgkg^{-1} and either 10 or 40 Jcm^{-2}; the third group of mice received only Formulation 2; the fourth and fifth groups of mice received only the energy dose of 10 or 40 Jcm^{-2} and the last group did not get any injection.

	Formulation 2	Energy dose	Number of mice
Group 1	+	+ (10J/cm²)	5
Group 2	+	+ (40J/cm²)	5
Group 3	+	-	5
Group 4	-	+ (10J/cm²)	5
Group 5	-	+ (40J/cm²)	5
Group 6	-	-	5

Table 4.1: Scheme with the description of the 6 groups of mice used in the first PDT experiment.

To analyse the effect of the PDT treatment on this OS mouse model we, qualitatively, evaluated the amount of lytic lesions present in the x-rays, calculated the primary tumour volume and counted the number of metastases.

No effect was visible in the four negative control groups demonstrating that the single elements (Formulation 2 or energy dose) do not have any effect on their own.

Surprisingly the treatment induced strong side effects and complications in the mice treated with 40Jcm^{-2} and 0.15mgkg^{-1} therefore we had to sacrifice the mice the day after performing the treatment (day 15) and they were not included in our further evaluation.

The other treated group of mice (0.15mgkg^{-1} plus 10Jcm^{-2}) showed a strong swelling with a starting process of necrosis at the level of the leg where the illumination was performed within 1 week from the treatment (day 21) therefore we decided to finish the experiment at this time point.

In the four negative controls groups tumour growth was comparable and no significant difference was noticed in the tumour volume; instead, the treated group (Formulation 2 plus 10Jcm^{-2}) shows a non-significant but larger tumour volume compared to all the other groups which is provoked by the swelling process that was induced by PDT (Fig. 4.13).

Since this process did not regress and the legs started to be necrotic we decided to repeat the experiment and change the treatment protocol by reducing the concentration of Formulation 2.

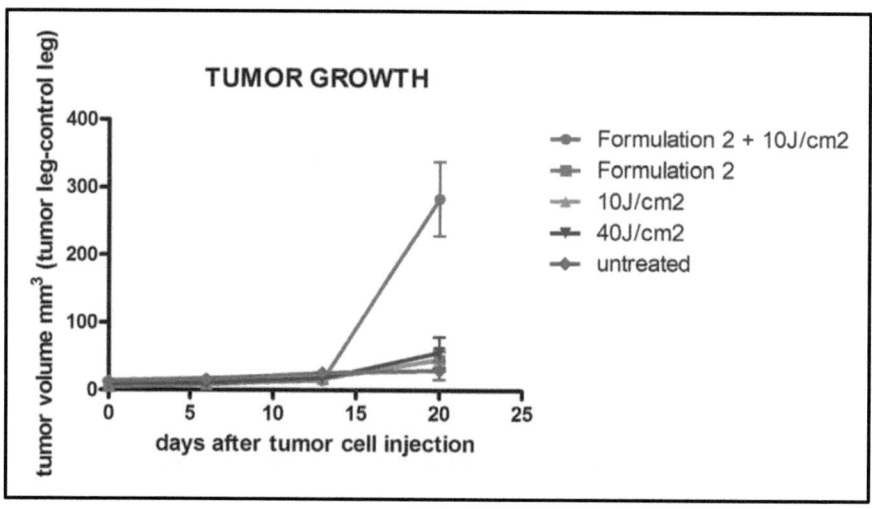

Fig.4.13: Tumour growth curve of the different groups of 143B*LacZ* tumour bearing mice: one group received Formulation 2 and energy dose (10Jcm^{-2}); one group received only Formulation 2; the other two groups received only energy dose (either 10Jcm^{-2} or 40Jcm^{-2}) and the last group received neither Formulation 2 nor energy dose.

X-ray pictures do not show any visible difference in the extent of lytic lesions in the left tibia of the mice (Fig. 4.14).

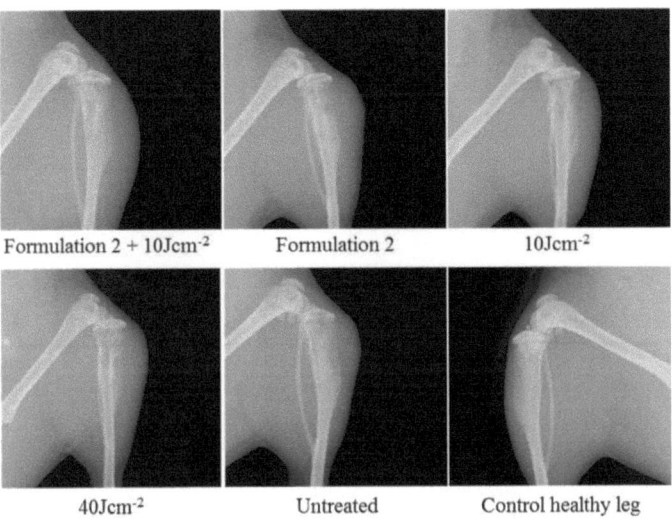

Fig.4.14: Representative x-ray pictures of a mouse from each different group: the x-ray pictures were taken at day 20 after tumour cell injection. The last picture is from the control non injected healthy leg for comparison.

Interestingly macro-metastases (diameter > 0.2mm) did not significantly differ in the five groups while micro-metastases (diameter < 0.2mm) were significantly lower in the treated group compared to two negative control groups: the group that received only Formulation 2 or only the energy dose of $10 Jcm^{-2}$ ($p<0.05$). The difference in the number of micro-metastases was not significantly different comparing the treated group with the untreated one because of high heterogeneity in the last mentioned group ($p=0.1023$) and close to significance compared to the group that received only energy dose ($40\ Jcm^{-2}$) ($p=0.0615$) (Fig. 4.15).

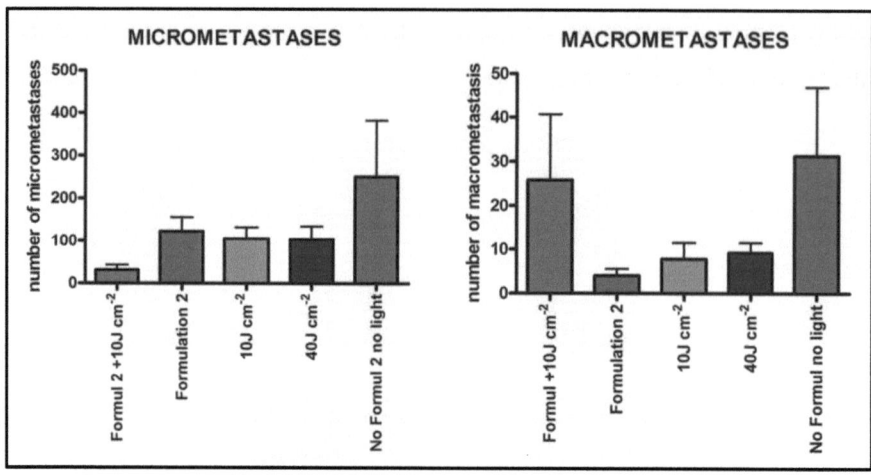

Fig.4.15: Graphs showing the number of micro- (diameter < 0.2mm) and macro-metastases (diameter > 0.2mm) in the five different groups.

In summary we conclude that the PDT protocol used in this first experiment has to be further optimised and we can assess that if PS and energy dose are not combined no effect can be visualised since the negative controls group did not show any difference in tumour growth, osteolysis and number of metastases.

In order to find an optimal concentration of Formulation 2 that would have an effect without causing strong effect in the mice we designed a second treatment study. In this experiment we kept the energy dose of $10Jcm^{-2}$ and reduced the Formulation 2 concentration to 0.05, 0.015 and $0.005 mgkg^{-1}$. As a negative control mice received only the highest dose of Formulation 2 among the ones used for the treatment ($0.05 mgkg^{-1}$) (Table 4.2).

As for the first experiment, also in this experiment the treatment was started at day 14 when lytic lesions could be visualised by x-ray in all the mice.

	Formulation 2 (mg kg⁻¹)	Energy dose	Number of mice
Group 1	0.05	+ (10J/cm²)	5
Group 2	0.015	+ (10J/cm²)	6
Group 3	0.005	+ (10J/cm²)	5
Group 4	0.05	-	5

Table 4.2: Scheme with the description of the 4 groups of mice used in the second PDT experiment.

The tumour volume was comparable in 3 of the 4 groups. In the group where mice received 0.015mgkg⁻¹, the mice showed in tendency a higher tumour volume which is not statistically significant and supposingly not dependent on the treatment (Fig. 4.16).

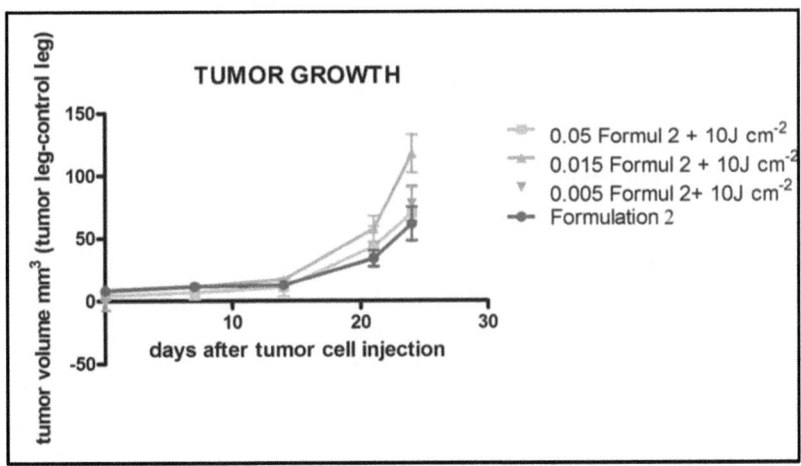

Fig.4.16: Tumour growth curve of the different groups of 143B*LacZ* tumour bearing mice: either they received Formulation 2 with a specific dose (0.05, 0.015 or 0.005mgkg⁻¹) and energy dose of 10Jcm⁻² or only 0.05mgkg⁻¹ of Formulation 2 (negative control).

The x-rays show that the mice treated with 0.05mgkg⁻¹ and an energy dose of 10Jcm⁻² have a smaller extent of osteolysis around the tibia compared to the mice from the other three groups: the control non injected healthy leg is shown to evaluate the degree of the lysis as non-lytic bone (Fig. 4.17). Concerning the metastases we could not detect significant differences among the groups (Fig.4.18).

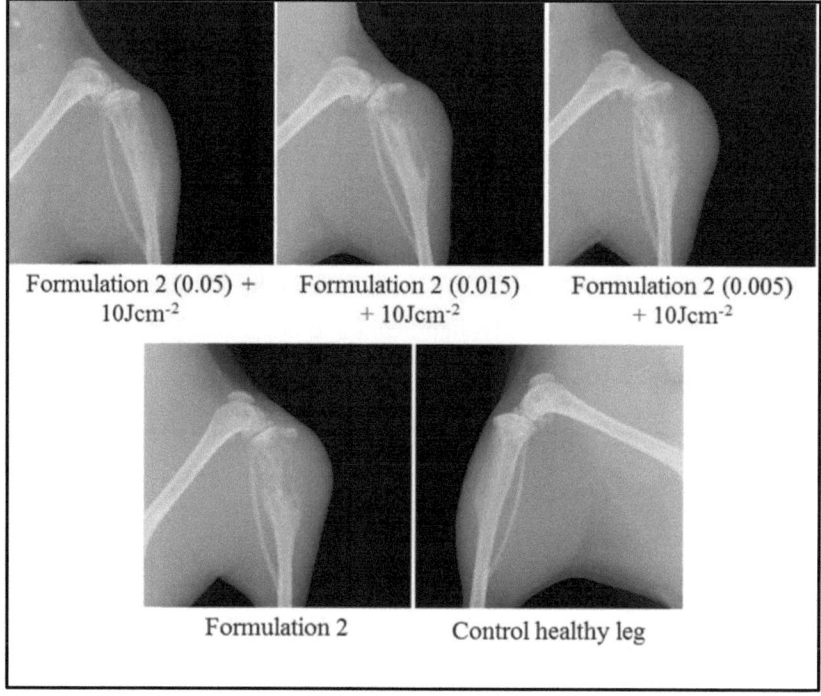

Fig.4.17: Representative x-ray pictures of a mouse from each different group: the x-ray pictures were taken at day 20 after tumour cell injection. The last picture is from the control non injected healthy leg for comparison.

Chapter 4 Evaluation of efficacy of Photodynamic Therapy in OS

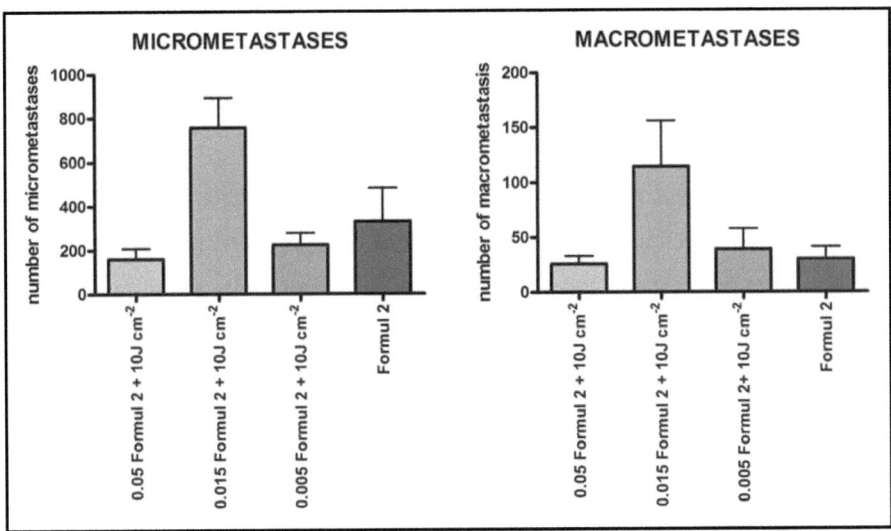

Fig.4.18: Graphs showing the number of micro- (diameter < 0.2mm) and macro-metastases (diameter > 0.2mm) in the five different groups.

4.2 DISCUSSION & OUTLOOK

The focus of this study was the evaluation of the efficacy of PDT in OS *in vitro* and *in vivo*.

In vitro we investigated two different OS cell line systems consisting of the highly metastatic human 143B and MG63-M8 cell lines and their respective low metastatic parental cell lines HOS and MG63. Importantly, we showed that the highly metastatic 143B and MG63-M8 OS cells take up more PS than the low metastatic parental HOS and MG63 cells (Reidy, Campanile et al. 2012). To our knowledge, previous studies exclusively compared the uptake of PS in tumour tissue with that in normal tissues, but the uptake by low and highly metastatic tumour cells has so far not been compared. Differences in uptake between normal and tumour tissue have been explained by higher proliferation rates of tumour compared to normal cells, by a high permeability or leakage of tumour blood vessel, low pH values in tumour tissue and by non-functional lymphatic drainage (Jori 1996; Juzeniene, Peng et al. 2007). If active uptake of PS is considered, the differences in uptake between low and highly metastatic tumour cells may be explained by different levels of transporter protein expression. Such mechanisms have not been examined here and need to be investigated in future studies.

In preparation for future investigations *in vivo*, additional *in vitro* experiments focused exclusively on the 143B cell line that, upon injection into the tibia of SCID mice, precisely reproduces the subtype of the human disease with a primary osteolytic bone tumour that predominantly metastasises to the lung.

In 143B cells, we first examined the subcellular localisation of PS. Confocal microscopy revealed perinuclear accumulation of the PS observed over time. This finding is consistent with that of Lassalle et al., who observed in HeLa cells accumulation of PS also preferentially in the perinuclear region (Lassalle, Wagner et al. 2008). In contrast, another study in MCF-7 cells showed a selective uptake in the endoplasmic reticulum and in the Golgi apparatus (Teiten, Bezdetnaya et al. 2003). Moreover, in human nasopharyngeal carcinoma and murine myeloid leukemia cells the PS was found localised predominantly in the mitochondria and in colon carcinoma PS was observed in lysosomes (Kessel 1999; Yow, Chen et al. 2000; Leung, Sun et al. 2002). Interestingly, these differences in subcellular accumulation had no effect on PDT efficacy in the different cell types (Reidy, Campanile et al. 2012).

Dark- and photo-toxicity of the PS mTHPC used in our studies in 143B cells were observed at concentrations $\geq 2.5 \mu gml^{-1}$ and $\geq 0.075 \mu gml^{-1}$, respectively, upon incubation of the cells with PS for 5 hrs. Previous studies in gall bladder and bile duct cancer cells revealed dark-toxicity at $10\text{-}100 \mu gml^{-1}$ PS after 20hrs of incubation. This suggests that 143B cells are more sensitive to PDT than other tumour cell lines (Kiesslich, Berlanda et al. 2007). This is also reflected by the photo-toxic potency of mTHPC in 143B cells. The LD_{50} of mTHPC at a laser-light dose of $2.5 Jcm^{-2}$ was $0.047 \mu gml^{-1}$ (±0.007 SEM), which is 10 times less than what was reported for human colon carcinoma (LD_{50} $0.45 \pm 0.15 \mu gml^{-1}$ at $3 Jcm^{-2}$) (Leung, Sun et al. 2002; Reidy, Campanile et al. 2012). A similar result was obtained in gall bladder and bile duct cancer cells with the same formulation Formulation 2 used in our study. The respective LD_{50} were 0.061 and $0.056 \mu gml^{-1}$ at $1.5 Jcm^{-2}$ illumination (Kiesslich, Berlanda et al. 2007). Of note, the study with the human colon carcinoma cells was performed with a different mTHPC formulation (Foscan®) and an incubation time of 20hrs, which can definitively influence the results. Indeed, the liposomal formulation increased the speed of PS uptake in the tumour cells since they express high levels of lipoproteins in the plasma membrane (Guelluy, Fontaine-Aupart et al. 2010). Taken together, the high sensitivity of 143B cells to mTHPC-mediated PDT is promising for future treatment studies *in vivo*.

In view of these promising results, it was of particular interest to investigate the mechanisms of cell death provoked by PDT, which apparently differs in different cell types. In 143B cells we observed caspase-dependent apoptosis already at 90min after activation of the PS laser-light illumination. Other studies showed the same result. Apoptosis rather than necrosis was an important finding because necrosis induces a stronger inflammatory response than apoptosis and consequently more severe side effects (Leung, Sun et al. 2002; Oleinick, Morris et al. 2002; Juarranz, Jaen et al. 2008).

It is well established that apoptosis in response to PDT frequently occurs in cancer in which PS is localised in the mitochondria. Autophagy, on the other hand, is believed to occur primarily in cells that accumulate PS in late endosomes, the endoplasmic reticulum and mitochondria. Autophagy has also been observed in cells expressing defective apoptosis signaling molecules. This has been demonstrated in the human mammary carcinoma cell line MCF with a mutation affecting the activity of caspase-3 (Guelluy, Fontaine-Aupart et al. 2010; Reiners, Agostinis et al. 2010). In summary, the results of the here presented *in vitro* studies justify investigations in *vivo* to optimise PDT protocols in orthotopic OS mouse models that reproduce the human disease.

The aim of a first series of experiments was the evaluation of the uptake of the liposomal formulation Formulation 2 of mTHPC and of the effects of illumination on primary tumour development and lung metastasis in order to move a step forward towards an intraoperative application of PDT in OS.

In the 143B cell-derived intratibial OS model in SCID mice, the uptake of intravenously administered Formulation 2 peaked at between 6 and 24hrs. A similar study performed in a mouse model of mammary sarcoma showed maximal uptake of Formulation 2 at between 6 and 15hrs, which is in good agreement with our study (Lassalle, Dumas et al. 2009). Comparable results were also found when an mTHPC formulation based on pegylated liposomes was used in a mouse model of mammary tumours. The peak of PS uptake in the tumour was reached 15hrs after PS injection. Foscan®, a formulation of mTHPC, in ethanol and propylene glycol, on the other hand, showed a maximal uptake by tumour tissue at between 24 and 48hrs after PS injection (Lassalle, Dumas et al. 2009). These findings pointed to a faster uptake of liposomal and pegylated liposomal formulations of mTHPC by tumour cells. Along these lines it was suggested that the pegylation does not increase the amount of internalised PS into the cell, but, as described previously, increases the stability of the molecule in the circulation by reducing the kidney clearance and interactions with cellular membrane proteins (Harris and Chess 2003).

A next series of here reported experiments aimed at optimising a PDT treatment efficacy in our orthotopic OS mouse model. Previous PDT studies in experimental tumour models in mice were exclusively carried out in subcutaneous models addressing questions concerning PS bio distribution and effects of tumour oxygenation on the success of PDT treatment (Chen, Huang et al. 2002; D'Hallewin, Kochetkov et al. 2008; Lassalle, Dumas et al. 2009). In these studies, side effects of PDT treatment, such as damage to healthy organs, tissues and blood vessels close to the tumour were not considered. A comparison between the results presented here and those of studies with subcutaneous models showed that the treatment conditions need to be redefined in orthotopic models. PS doses tolerated and not causing severe side effects due to damage of healthy tissue in our intratibial OS mouse model amounted to between 0.05 and 0.1mgkg^{-1} mTHPC, whereas in subcutaneous models routinely used mTHPC doses were between 0.15 and 0.3mgkg^{-1}. Mice with intratibial OS, treated in the here reported study with 0.15mgkg^{-1} mTHPC, showed a strong inflammatory response, reflected by tumour leg swelling, and areas of necrotic skin and subjacent tissue were observed in the illuminated leg.

The PDT treatment efficacy was also remarkably low. Despite a reduction in lytic lesions recognized by X-ray in treated compared to control animals, the primary tumour volumes were indistinguishable. A significant reduction in the number of pulmonary micro-metastases was only observed in the group of mice that received 0.15mgkg^{-1} mTHPC and, consequently, as outlined above, developed serious adverse effects (necrosis and swelling) in the tumour leg.

Plausible explanations for the low efficacy of PDT in our study are provided by a paper of Korbelik et al. (Korbelik, Krosl et al. 1996). This study compared tumour responses to PDT in BALB/c, nude and SCID mice. Interestingly, PDT was curative in the BALB/c mice but not in the nude and SCID mice. These findings suggested that the lack of a fully functional immune system in nude and SCID mice impaired PDT efficacy. This interpretation of the data was further supported by the observations that only bone marrow transplantations and not only a T-cell transfer rescued the PDT efficacy in immune compromised such that it became comparable to that observed in immunocompetent BALB/c mice. From all these observations, we conclude that effective PDT is dependent on a two-phase activation of the immune system. First, PDT activates the migration of myeloid inflammatory cells to the tumour and second it triggers a long lasting T-cell reaction (de Vree, Essers et al. 1996; Korbelik, Krosl et al. 1996).

Thus, additional studies in a syngeneic mouse OS model providing an intact immune system are required, but the here presented study revealed helpful data for a final optimisation of a PDT protocol for the treatment of intratibial OS in mice.

Additional experiments are needed in order to further evaluate the potential of PDT for intraoperative treatment of OS in patients.

Based on the here presented result, PDT experiments need to be done in animals with an intact immune system that sustains the treatment effect. The SCID mice lack a T-cell response and therefore cannot reproduce an adaptive immune response, which is needed for a fully effective PDT. Thus, and OS model in mice with an intact immune system that fully reproduces human OS is needed. Unfortunately, the syngeneic intratibial LM8 OS that has been established in the laboratory metastasises to the ovaries, the kidney, the liver and the lung, a pattern that is not observed in OS patients. Accordingly, we are currently in the process of establishing two new orthotopic syngeneic mouse models with the parental low metastatic mouse OS cell line K12 and the highly metastatic K7M2 derivative thereof with these two models we aim at reproducing predominant lung metastasis in experimental OS.

A success-limiting property of PDT in experimental and clinical applications is the remarkably low selective uptake of PS by tumour compared to healthy tissue. The uptake experiments performed in this thesis showed that the tumour leg takes up more PS than the liver or lung. The liposomal formulation that we used in the here presented study is considered to improve the selective uptake of the PS by tumour tissue. However, in recent years, new efforts have been made to conjugate PS to antibodies selective for tumour cell surface proteins in order to more specifically target tumour cells for PDT. Together with the only local illumination of tumour affected tissue, this considerably enhances the tumour selectivity of the treatment. In collaboration with Prof. Patrick Boisseau (Grenoble, France) and with biolitec research GmbH (Jena, Germany) a new project was started in 2010 aiming at the generation of mTHPC containing nanoparticles decorated with antibodies to CXCR4, which is abundantly expressed on the surface of some OS tumour cell lines. Depending on the results obtained with antibody-coated-nanoparticles, the next step is to aim for a clinical application.

4.3 MATERIAL & METHODS

4.3.1 MATERIAL AND OS CELL LINES

Formulation 1 and Formulation 2, two liposomal formulations of the PS called 5,10,15,20-tetrakis(meta-hydroxyphenyl)chlorine (mTHPC), were kindly provided by biolitec research GmbH, Jena, Germany in a 9:1 mixture of dipalmitoylphosphatidylcholine (DPPC) and dipalmitoylphosphatidylglycerol (DPPG) (> 99% purity). Formulation 1 was finally dissolved in water while Formulation 2 in 5% with a final concentration of 1.5mgml^{-1} mTHPC. Human OS low metastatic parental HOS (CRL 1543) cells were obtained from the American Type Culture Collection, Rockville, MD while the highly metastatic derivative 143B cells were purchased from ECACC (Salisbury, UK). HOS and 143B cells used in this study were stably transduced with a *LacZ* gene (143B*LacZ*, HOS*LacZ* cells), which will be needed in future *in vivo* studies for post-mortem visualisation of X-gal stained tumor tissue and metastases in mouse models at high sensitivity and resolution as reported previously (Arlt, Banke et al. 2011). Human OS low metastatic parental MG63 cells were kindly provided by Dr. G. Sarkar (Mayo Clinic, Rochester, MN, USA) while the highly metastatic derivative MG63-M8 by Dr. W.T. Zhu (Tongji Hospital, Huazhong University of Science and Technology, Wuhan, China): both cell lines were used only *in vitro* to confirm the uptake results in two different cell systems, therefore they are not *LacZ* tagged. MG63 and MG63-M8 were cultured in a humidified atmosphere of 5% CO_2 in DMEM (4.5g/l glucose)/Ham F12 (1:1) supplemented with 10% heat-inactivated FCS (cell culture medium). 143B*LacZ* and HOS*LacZ* were cultured in the same medium described above with the addition of 1200μgml^{-1} G418 for selection for neomycin resistance and stable expression of the *LacZ* gene. Experiments were also performed under these conditions. The pan-caspase inhibitor Z-VAD-FMK (BD Pharmingen AG,

Allschwil, Switzerland) was administered to the cells at 50 or 100μM 1hr prior to illumination until harvesting.

4.3.2 UPTAKE OF LIPOSOMAL FORMULATION 1

All the cell lines were seeded in 96-well plates (10'000 per well) and incubated at different time points (from 0 to 24hrs) with 2.5μgml^{-1}. After each time point the fluorescence intensity was measured at 652nm after excitation at 452nm with a Spectramax Gemini XS plate reader (Molecular Devices, Sunnyvale, CA) and the values from 143B*LacZ* and MG63 in the individual graphs were normalised to that of the respective parental, low metastatic cell lines HOS*LacZ* and MG63-M8 at 24hrs (100%).

In order to get a fluorescent value that would take into account the diameter and the number of the cells HOS*LacZ* and 143B*LacZ* cells were seeded in 6-well plates (0.2 x 10^6 cells per well) and allowed to adhere overnight. They were then incubated with the same concentration of Formulation 1 (0.6μgml^{-1}) for 2.5, 5 or 10hrs or with indicated concentrations of Formulation 1 (0-10μgml^{-1}) for 5hrs in the dark. The medium was removed and the cells were washed three times with PBS. They were then detached with trypsin/EDTA, centrifuged at 400 x g for 5min, washed twice with PBS, collected by centrifugation and the cell pellets were resuspended in PBS. Intracellularly accumulated Formulation 1 was quantified by measuring the fluorescence of 20'000 cells in 100μl PBS at 652nm with excitation at 417nm with a Spectramax Gemini XS plate reader (Molecular Devices, Sunnyvale, CA). A standard curve with known dilutions of Formulation 1 in 100μl PBS was used for quantification. The mean diameter of cells in suspension was calculated from 20 trypsinized cells with a Zeiss Observer.Z1 microscope (Axio Observer, Axio Vision Release 4.6.3 SP1, Jena, Germany) and the cell volume calculated with the equation for spheres. Formulation 1 fluorescence was then normalised to the cell volume. Subcellular localisation of Formulation 1by confocal laser scanning microscopy (CLSM) was carried out with 30`000 cells per well seeded on coverslips in 24-well plates and grown for 48hrs. The cells were incubated at 37°C for 5hrs in cell culture medium containing indicated concentrations of Formulation 1. Subsequent manipulations were all carried out at room temperature. The cells were first washed once with PBS and then fixed with 4% paraformaldehyde for 10min and washed three times with PBS. Cells were then preincubated in PBS/0.1% BSA/0.1% saponin for 10min and subsequently stained with 0.3nM DAPI in PBS/0.1% BSA/0.1% saponin for 15min in the dark. After another 3 washing steps with PBS/0.1% BSA/0.1% saponin, cover slips were briefly rinsed with tap water and mounted with Immumount (Thermo Shandon, Pittsburg, PA). CLSM was done with a Leica SP5 confocal microscope (Leica, Heidelberg, Germany; Center of Microscopy at the University of Zurich). Fluorescent probes were excited at 405nm (DAPI) and 415 nm (Formulation 1) and emitted light collected at 435-468nm for DAPI and at 591-705nm for Formulation 1. CLSM images of 1024 x 1024 pixel size in 8 bit were recorded under an oil immersion objective with a 63-fold magnification and a numerical aperture of 1.4. The pinhole diameter was set to 1 Airy unit with a size of 114.74µm. Images were processed with Leica LCS Lite software.

4.3.3 CYTOTOXICITY ASSAY

143B cells were seeded at a density of 3'000 cells per well in 96-well plates and allowed to adhere overnight. The cells were incubated with Formulation 1 in cell culture medium at indicated concentrations for 5hrs in the dark. The medium was then removed and the cells washed twice with 100µl PBS before 100µl fresh medium were added. The cells were then either kept in the dark or illuminated with a 652nm diode laser (Applied Optronics Corp., South Plainfield, NJ), equipped with a fiber-based frontal light distributor (Medlight SA, Ecublens, Switzerland). Energy dose of 2.5 Jcm^{-2}, 5 Jcm^{-2} or 10 Jcm^{-2} was applied at a fluence rate of 11.3$mWcm^{-2}$. Dark- and photo-toxicity were assessed

24hrs later by measuring cell metabolic activity with the water-soluble tetrazolium (WST-1) reagent (Roche Diagnostics AG, Rotkreuz, Switzerland). The % of viability was calculated as the ratio of absorbance at 415 nm of extracts of treated and non-treated cells multiplied by 100. For the determination of the number of 143B cells after PDT treatment, cells were seeded at a density of 20'000 cells per well in 24 well plates and allowed to adhere overnight. They were then incubated with Formulation 1 at indicated concentrations for 5hrs and subsequently illuminated with $5 Jcm^{-2}$ as described above. 24hrs later the cell culture medium was collected and the cells were washed once with PBS and then trypsinised. The collected medium, the PBS wash fraction and the trypsinised cells were pooled and centrifuged at 400 x g for 5 min. The supernatant was discarded and the cells resuspended in 200µl fresh medium. The cell suspension was transferred to a 96-well plate and the cells counted with a Guava EasyCyte machine (Guava Technologies Inc.).

4.3.4 WESTERN BLOT ANALYSIS

2.5×10^6 143B cells were seeded in 10 cm dishes and allowed to adhere overnight. They were then incubated with Formulation 1 at 0, 0.03, 0.15, 0.6, $1.25 \mu gml^{-1}$ for 5hrs and subsequently illuminated with $5 Jcm^{-2}$ as described above. Ninety min after illumination, the cells were washed three times with PBS/0.05% EDTA on ice. Washed cells and cells in the wash fractions were collected in one pellet. Cell pellets were lysed in 50mM tris-hydroxyethylaminomethan, pH 7.5, 150mM NaCl, 1% NP-40, 0.5% deoxycholic acid (DOC), 0.1% SDS, 1mM PMSF, sonicated and incubated at 4°C for 1 hr. Cell lysates were cleared by centrifugation at 20'800 x g and 4°C for 20min. Equal amounts of protein were loaded and separated by SDS-PAGE in 15% gels for cleaved caspase 3, in 12% gels for cleaved caspase 7 and full length and cleaved caspase 9 and in 8% gels for full-length and cleaved PARP. Proteins were transferred to nitrocellulose membranes by semi-dry blotting.

Cleaved caspase 3 and 7 and intact and cleaved caspase 9 as well as intact and cleaved poly-ADP-ribose polymerase (PARP) were detected with respective antibodies (Cell Signaling Technology, MA, USA). β-actin, used as a protein loading control, was detected with β-actin antibodies (Chemicon, Dietikon, Switzerland). Horseradish peroxidase (HRP)-tagged antibodies were obtained from Santa Cruz Biotechnology (Santa Cruz, CA, USA) and immunostained proteins were visualised by detection of HRP substrate Immobilon (Millipore, Billerica, MA) chemoluminescence with a VersaDoc^{TM} Imaging System (Bio-Rad, Munich, Germany).

4.3.5 MOUSE MODELS

Animal care and experiments were conducted in accordance with Swiss Animal Welfare legislation and were approved by the Veterinary Office of Canton Zurich, Zurich, Switzerland. Six to eight weeks old female immunosuppressed SCID and immunocompetent C3H mice were purchased from Charles River Laboratories (Sulzfeld, Germany), kept in pathogen-free conditions. The mice have always reached the mouse facility at least 1 week before starting any experiment. Before injection, cells were detached with trypsin-ethylenediaminetetraacetic acid (EDTA), washed twice with PBS containing 0.05% EDTA and finally re-suspended in PBS-EDTA at a final concentration of 5×10^7 cellsml^{-1}. An aliquot of the cell suspension (10µl) was orthotopically injected into the medullar cavity of the left tibia of the mice. Primary tumour development was monitored by X-ray (Faxitron, USA) weekly for 143B and LM8 cell-derived tumours.

Primary tumour volume was measured with a caliper along the length and the width of the leg and calculated according to the formula: $(L \times W^2)/2$. With the same formula the volume in the control healthy leg was calculated and subtracted to the volume in the tumour leg to exclude from the tumour volume the part coming from the bone and from the physiological growth of the mice.

The 143B*LacZ* injected mice were sacrificed between days 21 and 23 after tumour cell inoculation; LM8*LacZ* mice between day 21-28.

Sacrifice was performed using an optimised protocol which is specific for *LacZ* tagged cells. After anaesthetising the mice 10% Ketamin, 2% Xylazin and 10% Acepromazin, the mice were opened and lungs were first perfused through the right ventricle with PBS 1x and then with PFA 3% from the right ventricle and from the trachea. While the lungs have been fixed, the primary tumour and the healthy control leg were removed. The last step involved the excision of the lungs which were, first, washed with PBS 1x and finally excised. Primary tumour and lungs were then kept in fixing solution (PFA 2%) for 30 min, the organs were stained with X-Gal staining solution for at least 3hrs and after washing them 3 times with PBS 1x they were kept for long-term storage in 4% PFA (Arlt, Banke et al. 2011).

Micro- (diameter larger than 0.2mm) and macro-metastases (diameter larger than 0.2mm) were counted on the surface of the whole lung using the Olympus SZX 10 microscope (Olympus, Southend on Sea, UK).

4.3.6 *IN VIVO* UPTAKE OF LIPOSOMAL FORMULATION 2

The *in vivo* uptake was measured with a PDT fluorometer package (JETI Technische Instrumente GmbH, Jena-Germany) provided by biolitec research GmbH containing the fluorometer and a measuring fiber. In the 143B*LacZ* model we evaluated the uptake in the tumour leg, in the liver which is the metabolic organ and in the lung where metastases appear in this model. Mice (n=3) were injected at day 21 after tumour cell injection intravenously with 1.5mgml^{-1} Formulation 2 and the uptake was measured *in vivo* with the optic fiber in contact with the tumour area. We measured the uptake at different time points (from 0 to 48hrs) at three different locations of each organ to get more reliable results and we used a mouse as the background for the fluorometer that was injected only with 5% glucose. The excitation of the violet LED lamp which is contained in the fluorometer is 405nm and the result is shown in a spectrum from 550 to 800 nm. The integration time for the measurement was kept fixed at 7500ms. Since Formulation 2 has the maximum emission peak at 652nm we used the relative fluorescence unit from this wavelength and calculated the average from the three measurements obtained from each mouse and from each organ.

In the experiment performed with the LM8*LacZ* model the procedure was the same but the Formulation 2 was injected in the mice at day 26 (n=2) in one experiment and at day 21 (n=7) in the second experiment since in the first experiment most of the mice had to be sacrificed before the experiment could be finished. Since in this model the metastases develop in liver, ovaries, kidney and lung we measured only the uptake in the tumour leg and in the healthy control leg to evaluate the specificity of the Formulation 2 uptake.

4.3.7 TREATMENT PROTOCOL

The therapy experiments were performed only with the 143B*LacZ* model and the therapy was started as soon as lytic lesions could be visualised by x-ray (day 14 after tumour cell injection).

In this study one group received nothing; one group received only $0.15 mgkg^{-1}$ of Formulation 2; a third and fourth group were only exposed to light dose ($10Jcm^{-2}$ or $40Jcm^{-2}$) and the fifth and sixth group received 0.15 $mgkg^{-1}$ Formulation 2 and later they received either $10Jcm^{-2}$ of energy dose or $40Jcm^{-2}$ energy dose. After irradiation mice were checked each day and one group of mice ($0.15mgkg^{-1}$ plus $40Jcm^{-2}$) had to be sacrificed already the day after the treatment because of strong side effects (necrosis in the leg, swelling, breathing affection and reduced motility). The experiment was concluded on the day 21 when also the second treated group ($0.15mgkg^{-1}$ plus $10Jcm^{-2}$) showed complications related to a strong inflammatory response in the tumour leg.

In the second pilot study the PS concentration were reduced and three dilutions were tested to 0.05, 0.015 and 0.005mgkg^{-1}. The negative control group in this experiment received only Formulation 2 (0.05mgkg^{-1}). No further control groups were used since no difference could be reported in tumour volume and number of metastases among the groups of negative control used in the first experiment. The mice were sacrificed at day 24.

The analysis of this experiment consisted in comparing the tumour volume among the groups, evaluating visually the osteolysis by x-ray and in counting the total number of micro- and macro-metastases.

CHAPTER 5

CONCLUSIONS

Principle aims of the studies reported in this thesis were the evaluation of metabolic imaging by PET and its prognostic power in experimental intratibial OS models, making use of clinically approved tracers for PET and an assessment of the potential of PDT for intraoperative OS treatment with the ultimate goal to improve long-term survival of OS patients.

Current routinely used non-invasive tumour imaging techniques, including x-ray, CT and MRI, are mainly diagnostic with little prognostic power, except when they are used to monitor the success of chemo- or radiotherapy, e.g. by CT. Monitoring of tumour treatment success is frequently performed with PET with ^{18}F-FDG as a tracer, which rapidly and rather selectively accumulates in tissues with a high glucose metabolism. However, in recent years, additional PET tracers have been developed for clinical use and are compounds that are substrates in distinct biological processes and consequently indicators of metabolic conditions in e.g. tumour tissue. The three distinct intratibial OS mouse models that we used for our PET studies individually represent distinct phenotypes of pathophysiological relevant conditions in OS tissue, including rapidly proliferating or hypoxic tumour domains and rapidly remodelled bone-like structures. The results presented in the thesis demonstrate that, with the use of ^{18}F-FDG, indicating high glucose metabolism and rapid proliferation, of ^{18}F-FMISO, indicating hypoxia, or of ^{18}F-Fluoride as an indicator of bone formation and remodelling, we were able to differentiate between pronounced and mild osteoblastic and osteolytic OS phenotypes, represented by the three experimental models investigated here. The detection of hypoxia in OS tumour tissue is of particular relevance because there is strong evidence for increased resistance of hypoxic OS tissue to chemotherapy with the consequence of a poor prognosis for the patient. All findings taken together, we can conclude that PET of OS tumours with a combination of the three metabolic tracers ^{18}F-FDG, ^{18}F-FMISO and ^{18}F-Fluoride has relevant prognostic potential in OS.

However, the presented results also demonstrate that we failed to detect even severe lung metastasis in the animal models investigated here with the PET scanner that was available to us. In order to take advantage of the also extremely important disease outcome predictor "metastasis" in OS, the signal to background ratio and the sensitivity of lung PET scans of mice need to be significantly improved. Efforts to achieve these improvements are worthwhile, because we were able to demonstrate selective uptake of the PET tracers and even calcification, indicated by ^{18}F-Fluoride uptake, in individual metastatic lesions of an osteoblastic model by *ex-vivo* autoradiography of lung tissue sections. Based on the results presented here and the fact that the PET tracers used are all clinically approved, prognostic PET imaging in OS patients can be envisaged. Prognostic differentiation of metabolic phenotypes of OS tumours by PET together with well-established histological grading of tumour biopsies will likely be of significant help for the design of effective, tumour phenotype-specific treatment regimens with the benefit of fewer side effects for the patients but maintained or even improved efficacy.

Along these lines, PDT, which is already clinically used for the treatment of epithelial tumours, is considered for intraoperative OS treatment to target rather frequently occurring OS-satellites in healthy tissue beyond tumour margins, which give rise to recurrences. The results of the experiments carried out in this thesis are encouraging. They demonstrate that an already clinically approved lipid formulation of the photosensitizer mTHPC is more efficiently taken up and accumulated by aggressive highly metastatic human 143B OS cells than by the parental low metastatic HOS cells. Moreover, 143B cells are more sensitive to dark- and photo-toxicity and undergo apoptosis rather than necrosis, which in living organisms will cause fewer side effects. The *in vivo* studies carried out so far in intratibial OS mouse models indicate that future experiments, required for a preclinical assessment of PDT as novel intraoperative treatment modality in OS, need to be performed in syngeneic intratibial OS mouse models with metastasis predominantly to the lung, because local immune responses in the tumour-affected tissue are crucial for maximising treatment efficacy.

In conclusion, the results of the studies presented in this thesis pave the way for further development of novel clinically applicable PET imaging strategies for early detection and prognosis of OS in patients and the design of more tumour-type tailored treatment. Based on the results obtained with PDT in our *in vitro* and *in vivo* experimental OS systems, there is considerable potential for intraoperative treatment of OS with this novel modality in combination with the well-established protocols of neo-adjuvant chemotherapy and, when required, radiotherapy.

6 REFERENCES

Abed, R. and R. Grimer (2010). "Surgical modalities in the treatment of bone sarcoma in children." Cancer Treat Rev **36**(4): 342-347.
Abudu, A., N. K. Sferopoulos, et al. (1996). "The surgical treatment and outcome of pathological fractures in localised osteosarcoma." J Bone Joint Surg Br **78**(5): 694-698.
Agostinis, P., K. Berg, et al. (2011). "Photodynamic therapy of cancer: an update." CA Cancer J Clin **61**(4): 250-281.
Akiyama, T., C. R. Dass, et al. (2008). "Novel therapeutic strategy for osteosarcoma targeting osteoclast differentiation, bone-resorbing activity, and apoptosis pathway." Mol Cancer Ther **7**(11): 3461-3469.
Aksnes, L. H., K. S. Hall, et al. (2006). "Management of high-grade bone sarcomas over two decades: the Norwegian Radium Hospital experience." Acta Oncol **45**(1): 38-46.
Alauddin, M. M., A. Shahinian, et al. (2001). "Preclinical evaluation of the penciclovir analog 9-(4-[(18)F]fluoro-3-hydroxymethylbutyl)guanine for in vivo measurement of suicide gene expression with PET." J Nucl Med **42**(11): 1682-1690.
Allison, R. R., G. H. Downie, et al. (2004). "Photosensitizers in clinical PDT." Photodiagnosis and Photodynamic Therapy **1**(1): 27-42.
Ametamey, S. M., M. Honer, et al. (2008). "Molecular imaging with PET." Chem Rev **108**(5): 1501-1516.
Arlt, M., C. Kopitz, et al. (2002). "Increase in gelatinase-specificity of matrix metalloproteinase inhibitors correlates with antimetastatic efficacy in a T-cell lymphoma model." Cancer Res **62**(19): 5543-5550.
Arlt, M. J., I. J. Banke, et al. (2011). "LacZ transgene expression in the subcutaneous Dunn/LM8 osteosarcoma mouse model allows for the identification of micrometastasis." J Orthop Res **29**(6): 938-946.
Bacci, G., D. Donati, et al. (1998). "[Local recurrence after surgical or surgical-chemotherapeutic treatment of osteosarcoma of the limbs. Incidence, risk factors and prognosis]." Minerva Chir **53**(7-8): 619-629.
Bacci, G., S. Ferrari, et al. (2003). "Nonmetastatic osteosarcoma of the extremity with pathologic fracture at presentation: local and systemic control by amputation or limb salvage after preoperative chemotherapy." Acta Orthop Scand **74**(4): 449-454.
Bacci, G., S. Ferrari, et al. (1994). "Prognostic significance of serum lactate dehydrogenase in patients with osteosarcoma of the extremities." J Chemother **6**(3): 204-210.
Bacci, G., A. Longhi, et al. (2006). "Prognostic factors for osteosarcoma of the extremity treated with neoadjuvant chemotherapy: 15-year experience in 789 patients treated at a single institution." Cancer **106**(5): 1154-1161.
Bacci, G., P. Picci, et al. (1993). "Prognostic significance of serum alkaline phosphatase measurements in patients with osteosarcoma treated with adjuvant or neoadjuvant chemotherapy." Cancer **71**(4): 1224-1230.
Bacci, G., M. Rocca, et al. (2008). "High grade osteosarcoma of the extremities with lung metastases at presentation: treatment with neoadjuvant chemotherapy and simultaneous resection of primary and metastatic lesions." J Surg Oncol **98**(6): 415-420.
Barwick, T., B. Bencherif, et al. (2009). "Molecular PET and PET/CT imaging of tumour cell proliferation using F-18 fluoro-L-thymidine: a comprehensive evaluation." Nucl Med Commun **30**(12): 908-917.
Bateson, E. M. (1965). "An Analysis of 155 Solitary Lung Lesions Illustrating the Differential Diagnosis of Mixed Tumours of the Lung." Clin Radiol **16**: 51-65.
Bellnier, D. A. (1991). "Potentiation of photodynamic therapy in mice with human tumor necrosis factor-alpha." J Photochem Photobiol B **8**(2): 203-210.
Belt, J. A., N. M. Marina, et al. (1993). "Nucleoside transport in normal and neoplastic cells." Adv Enzyme Regul **33**: 235-252.

Benz, M. R., J. Czernin, et al. (2009). "FDG-PET/CT imaging predicts histopathologic treatment responses after the initial cycle of neoadjuvant chemotherapy in high-grade soft-tissue sarcomas." Clin Cancer Res **15**(8): 2856-2863.

Bielack, S. S., D. Carrle, et al. (2008). "Bone tumors in adolescents and young adults." Curr Treat Options Oncol **9**(1): 67-80.

Bielack, S. S., B. Kempf-Bielack, et al. (2009). "Second and subsequent recurrences of osteosarcoma: presentation, treatment, and outcomes of 249 consecutive cooperative osteosarcoma study group patients." J Clin Oncol **27**(4): 557-565.

Bielack, S. S., B. Kempf-Bielack, et al. (2002). "Prognostic factors in high-grade osteosarcoma of the extremities or trunk: an analysis of 1,702 patients treated on neoadjuvant cooperative osteosarcoma study group protocols." J Clin Oncol **20**(3): 776-790.

Bieling, P., S. Bielack, et al. (1991). "[Neoadjuvant chemotherapy of osteosarcoma. Preliminary results of the cooperative COSS-86 osteosarcoma study]." Klin Padiatr **203**(4): 220-230.

Biermann, J. S., D. R. Adkins, et al. (2010). "Bone cancer." J Natl Compr Canc Netw **8**(6): 688-712.

Borst, P., R. Evers, et al. (2000). "A family of drug transporters: the multidrug resistance-associated proteins." J Natl Cancer Inst **92**(16): 1295-1302.

Brahimi-Horn, C., E. Berra, et al. (2001). "Hypoxia: the tumor's gateway to progression along the angiogenic pathway." Trends Cell Biol **11**(11): S32-36.

Brenner, W., K. H. Bohuslavizki, et al. (2003). "PET imaging of osteosarcoma." J Nucl Med **44**(6): 930-942.

Bruehlmeier, M., B. Kaser-Hotz, et al. (2005). "Measurement of tumor hypoxia in spontaneous canine sarcomas." Vet Radiol Ultrasound **46**(4): 348-354.

Buck, A. K., K. Herrmann, et al. (2008). "Imaging bone and soft tissue tumors with the proliferation marker [18F]fluorodeoxythymidine." Clin Cancer Res **14**(10): 2970-2977.

Burch, S., C. London, et al. (2009). "Treatment of canine osseous tumors with photodynamic therapy: a pilot study." Clin Orthop Relat Res **467**(4): 1028-1034.

Caroli, P., C. Nanni, et al. (2010). "Non-FDG PET in the practice of oncology." Indian J Cancer **47**(2): 120-125.

Carrle, D. and S. S. Bielack (2006). "Current strategies of chemotherapy in osteosarcoma." Int Orthop **30**(6): 445-451.

Castano, A. P., P. Mroz, et al. (2006). "Photodynamic therapy and anti-tumour immunity." Nat Rev Cancer **6**(7): 535-545.

Chen, B., B. W. Pogue, et al. (2003). "Blood flow dynamics after photodynamic therapy with verteporfin in the RIF-1 tumor." Radiat Res **160**(4): 452-459.

Chen, Q., Z. Huang, et al. (2002). "Improvement of tumor response by manipulation of tumor oxygenation during photodynamic therapy." Photochem Photobiol **76**(2): 197-203.

Chou, A. J. and R. Gorlick (2006). "Chemotherapy resistance in osteosarcoma: current challenges and future directions." Expert Rev Anticancer Ther **6**(7): 1075-1085.

Clark, J. C., C. R. Dass, et al. (2008). "A review of clinical and molecular prognostic factors in osteosarcoma." J Cancer Res Clin Oncol **134**(3): 281-297.

Cosse, J. P. and C. Michiels (2008). "Tumour hypoxia affects the responsiveness of cancer cells to chemotherapy and promotes cancer progression." Anticancer Agents Med Chem **8**(7): 790-797.

Costelloe, C. M., H. A. Macapinlac, et al. (2009). "18F-FDG PET/CT as an indicator of progression-free and overall survival in osteosarcoma." J Nucl Med **50**(3): 340-347.

Cramers, P., M. Ruevekamp, et al. (2003). "Foscan uptake and tissue distribution in relation to photodynamic efficacy." Br J Cancer **88**(2): 283-290.

Culverwell, A. D., A. F. Scarsbrook, et al. (2011). "False-positive uptake on 2-[(1)F]-fluoro-2-deoxy-D-glucose (FDG) positron-emission tomography/computed tomography (PET/CT) in oncological imaging." Clin Radiol **66**(4): 366-382.

D'Hallewin, M. A., D. Kochetkov, et al. (2008). "Photodynamic therapy with intratumoral administration of Lipid-Based mTHPC in a model of breast cancer recurrence." Lasers Surg Med **40**(8): 543-549.

Dai, C. Y., C. M. Haqq, et al. (2006). "Molecular correlates of site-specific metastasis." Semin Radiat Oncol **16**(2): 102-110.

Davis, S. D. (1991). "CT evaluation for pulmonary metastases in patients with extrathoracic malignancy." Radiology 180(1): 1-12.

de Vree, W. J., M. C. Essers, et al. (1996). "Evidence for an important role of neutrophils in the efficacy of photodynamic therapy in vivo." Cancer Res 56(13): 2908-2911.

Dehdashti, F., J. Picus, et al. (2005). "Positron tomographic assessment of androgen receptors in prostatic carcinoma." Eur J Nucl Med Mol Imaging 32(3): 344-350.

Deroose, C. M., A. De, et al. (2007). "Multimodality imaging of tumor xenografts and metastases in mice with combined small-animal PET, small-animal CT, and bioluminescence imaging." J Nucl Med 48(2): 295-303.

Diederichs, C. G., L. Staib, et al. (1998). "FDG PET: elevated plasma glucose reduces both uptake and detection rate of pancreatic malignancies." J Nucl Med 39(6): 1030-1033.

Dorfman, H. D. (1998). Bone Tumors. New York, Mosby.

Dubois, L., W. Landuyt, et al. (2004). "Evaluation of hypoxia in an experimental rat tumour model by [(18)F]fluoromisonidazole PET and immunohistochemistry." Br J Cancer 91(11): 1947-1954.

Duhaylongsod, F. G., V. J. Lowe, et al. (1995). "Detection of primary and recurrent lung cancer by means of F-18 fluorodeoxyglucose positron emission tomography (FDG PET)." J Thorac Cardiovasc Surg 110(1): 130-139; discussion 139-140.

Ebenhan, T., M. Honer, et al. (2009). "Comparison of [18F]-tracers in various experimental tumor models by PET imaging and identification of an early response biomarker for the novel microtubule stabilizer patupilone." Mol Imaging Biol 11(5): 308-321.

Ek, E. T., C. R. Dass, et al. (2006). "Commonly used mouse models of osteosarcoma." Crit Rev Oncol Hematol 60(1): 1-8.

Eppert, K., J. S. Wunder, et al. (2005). "von Willebrand factor expression in osteosarcoma metastasis." Mod Pathol 18(3): 388-397.

Even-Sapir, E., E. Mishani, et al. (2007). "18F-Fluoride positron emission tomography and positron emission tomography/computed tomography." Semin Nucl Med 37(6): 462-469.

Fass, L. (2008). "Imaging and cancer: a review." Mol Oncol 2(2): 115-152.

Fletcher, C. D. M. U., K.K. Mertens F. (2002). Pathology and Genetics of Tumours of Soft Tissue and Bone. W. H. Organization. Lyon, IARC*Press*.

Foukas, A. F., N. S. Deshmukh, et al. (2002). "Stage-IIB osteosarcomas around the knee. A study of MMP-9 in surviving tumour cells." J Bone Joint Surg Br 84(5): 706-711.

Franzius, C., S. Bielack, et al. (2002). "Prognostic significance of (18)F-FDG and (99m)Tc-methylene diphosphonate uptake in primary osteosarcoma." J Nucl Med 43(8): 1012-1017.

Franzius, C., H. E. Daldrup-Link, et al. (2001). "FDG-PET for detection of pulmonary metastases from malignant primary bone tumors: comparison with spiral CT." Ann Oncol 12(4): 479-486.

Franzius, C., M. Hotfilder, et al. (2006). "Successful high-resolution animal positron emission tomography of human Ewing tumours and their metastases in a murine xenograft model." Eur J Nucl Med Mol Imaging 33(12): 1432-1441.

Franzius, C., J. Sciuk, et al. (2000). "FDG-PET for detection of osseous metastases from malignant primary bone tumours: comparison with bone scintigraphy." Eur J Nucl Med 27(9): 1305-1311.

Friedman, M. A. and S. K. Carter (1972). "The therapy of osteogenic sarcoma: current status and thoughts for the future." J Surg Oncol 4(5): 482-510.

Fuchs, N., S. S. Bielack, et al. (1998). "Long-term results of the co-operative German-Austrian-Swiss osteosarcoma study group's protocol COSS-86 of intensive multidrug chemotherapy and surgery for osteosarcoma of the limbs." Ann Oncol 9(8): 893-899.

Gambhir, S. S., J. Czernin, et al. (2001). "A tabulated summary of the FDG PET literature." J Nucl Med 42(5 Suppl): 1S-93S.

Goorin, A. M., M. B. Harris, et al. (2002). "Phase II/III trial of etoposide and high-dose ifosfamide in newly diagnosed metastatic osteosarcoma: a pediatric oncology group trial." J Clin Oncol 20(2): 426-433.

Gopal, B. S. (2003). Fundamentals of Nuclear Pharmacy. New York, Springer.

Grosu, A. L., W. A. Weber, et al. (2005). "Reirradiation of recurrent high-grade gliomas using amino acid PET (SPECT)/CT/MRI image fusion to determine gross tumor volume for stereotactic fractionated radiotherapy." Int J Radiat Oncol Biol Phys 63(2): 511-519.

Grosu, A. L., W. A. Weber, et al. (2005). "L-(methyl-11C) methionine positron emission tomography for target delineation in resected high-grade gliomas before radiotherapy." Int J Radiat Oncol Biol Phys **63**(1): 64-74.

Guelluy, P. H., M. P. Fontaine-Aupart, et al. (2010). "Optimizing photodynamic therapy by liposomal formulation of the photosensitizer pyropheophorbide-a methyl ester: in vitro and ex vivo comparative biophysical investigations in a colon carcinoma cell line." Photochem Photobiol Sci **9**(9): 1252-1260.

Guise, T. A., K. S. Mohammad, et al. (2006). "Basic mechanisms responsible for osteolytic and osteoblastic bone metastases." Clin Cancer Res **12**(20 Pt 2): 6213s-6216s.

Gupta, G. P., D. X. Nguyen, et al. (2007). "Mediators of vascular remodelling co-opted for sequential steps in lung metastasis." Nature **446**(7137): 765-770.

Hacker, A., S. Jeschke, et al. (2006). "Detection of pelvic lymph node metastases in patients with clinically localized prostate cancer: comparison of [18F]fluorocholine positron emission tomography-computerized tomography and laparoscopic radioisotope guided sentinel lymph node dissection." J Urol **176**(5): 2014-2018; discussion 2018-2019.

Hanahan, D. and R. A. Weinberg (2011). "Hallmarks of cancer: the next generation." Cell **144**(5): 646-674.

Harris, J. M. and R. B. Chess (2003). "Effect of pegylation on pharmaceuticals." Nat Rev Drug Discov **2**(3): 214-221.

Heare, T., M. A. Hensley, et al. (2009). "Bone tumors: osteosarcoma and Ewing's sarcoma." Curr Opin Pediatr **21**(3): 365-372.

Helman, L. J. and P. Meltzer (2003). "Mechanisms of sarcoma development." Nat Rev Cancer **3**(9): 685-694.

Hicks, R. J., D. Rischin, et al. (2005). "Utility of FMISO PET in advanced head and neck cancer treated with chemoradiation incorporating a hypoxia-targeting chemotherapy agent." Eur J Nucl Med Mol Imaging **32**(12): 1384-1391.

Hogendoorn, P. C., N. Athanasou, et al. (2010). "Bone sarcomas: ESMO Clinical Practice Guidelines for diagnosis, treatment and follow-up." Ann Oncol **21 Suppl 5**: v204-213.

Honer, M., M. Bruhlmeier, et al. (2004). "Dynamic imaging of striatal D2 receptors in mice using quad-HIDAC PET." J Nucl Med **45**(3): 464-470.

Honer, M., B. Hengerer, et al. (2006). "Comparison of [18F]FDOPA, [18F]FMT and [18F]FECNT for imaging dopaminergic neurotransmission in mice." Nucl Med Biol **33**(5): 607-614.

Hong, H., Y. Zhang, et al. (2012). "In vivo targeting and positron emission tomography imaging of tumor vasculature with (66)Ga-labeled nano-graphene." Biomaterials **33**(16): 4147-4156.

Hu, G., R. A. Chong, et al. (2009). "MTDH activation by 8q22 genomic gain promotes chemoresistance and metastasis of poor-prognosis breast cancer." Cancer Cell **15**(1): 9-20.

Huvos, A. G. (1986). "Osteogenic sarcoma of bones and soft tissues in older persons. A clinicopathologic analysis of 117 patients older than 60 years." Cancer **57**(7): 1442-1449.

Iagaru, A., S. Chawla, et al. (2006). "18F-FDG PET and PET/CT for detection of pulmonary metastases from musculoskeletal sarcomas." Nucl Med Commun **27**(10): 795-802.

Imam, S. K. (2010). "Review of positron emission tomography tracers for imaging of tumor hypoxia." Cancer Biother Radiopharm **25**(3): 365-374.

Itoh, Y., M. Tamai, et al. (2002). "Involvement of multidrug resistance-associated protein 2 in in vivo cisplatin resistance of rat hepatoma AH66 cells." Anticancer Res **22**(3): 1649-1653.

Iwano, S., N. Makino, et al. (2004). "Solitary pulmonary nodules: optimal slice thickness of high-resolution CT in differentiating malignant from benign." Clin Imaging **28**(5): 322-328.

Jones, N. P. and A. Schulze (2012). "Targeting cancer metabolism--aiming at a tumour's sweet-spot." Drug Discov Today **17**(5-6): 232-241.

Jones, N. P. and A. Schulze (2012). "Targeting cancer metabolism - aiming at a tumour's sweet-spot." Drug Discov Today **17**(5-6): 232-241.

Jori, G. (1996). "Tumour photosensitizers: approaches to enhance the selectivity and efficiency of photodynamic therapy." J Photochem Photobiol B **36**(2): 87-93.

Josefsen, L. B. and R. W. Boyle (2008). "Photodynamic therapy and the development of metal-based photosensitisers." Met Based Drugs **2008**: 276109.

Juarranz, A., P. Jaen, et al. (2008). "Photodynamic therapy of cancer. Basic principles and applications." Clin Transl Oncol **10**(3): 148-154.
Juweid, M. E. and B. D. Cheson (2006). "Positron-emission tomography and assessment of cancer therapy." N Engl J Med **354**(5): 496-507.
Juzeniene, A., Q. Peng, et al. (2007). "Milestones in the development of photodynamic therapy and fluorescence diagnosis." Photochem Photobiol Sci **6**(12): 1234-1245.
Kang, Y., P. M. Siegel, et al. (2003). "A multigenic program mediating breast cancer metastasis to bone." Cancer Cell **3**(6): 537-549.
Kaste, S. C., C. B. Pratt, et al. (1999). "Metastases detected at the time of diagnosis of primary pediatric extremity osteosarcoma at diagnosis: imaging features." Cancer **86**(8): 1602-1608.
Kato, M., J. Kitayama, et al. (2003). "Expression pattern of CXC chemokine receptor-4 is correlated with lymph node metastasis in human invasive ductal carcinoma." Breast Cancer Res **5**(5): R144-150.
Kelloff, G. J., J. M. Hoffman, et al. (2005). "Progress and promise of FDG-PET imaging for cancer patient management and oncologic drug development." Clin Cancer Res **11**(8): 2785-2808.
Kenny, L., R. C. Coombes, et al. (2007). "Imaging early changes in proliferation at 1 week post chemotherapy: a pilot study in breast cancer patients with 3'-deoxy-3'-[18F]fluorothymidine positron emission tomography." Eur J Nucl Med Mol Imaging **34**(9): 1339-1347.
Kessel, D. (1999). "Transport and localisation of m-THPC in vitro." Int J Clin Pract **53**(4): 263-267.
Khanna, C., J. Khan, et al. (2001). "Metastasis-associated differences in gene expression in a murine model of osteosarcoma." Cancer Res **61**(9): 3750-3759.
Khanna, C., J. Prehn, et al. (2000). "An orthotopic model of murine osteosarcoma with clonally related variants differing in pulmonary metastatic potential." Clin Exp Metastasis **18**(3): 261-271.
Kido, A., T. Tsujiuchi, et al. (1997). "p53 mutation and absence of mdm2 amplification and Ki-ras mutation in 4-hydroxyamino quinoline 1-oxide induced transplantable osteosarcomas in rats." Cancer Lett **112**(1): 5-10.
Kiesslich, T., J. Berlanda, et al. (2007). "Comparative characterization of the efficiency and cellular pharmacokinetics of Foscan- and Foslip-based photodynamic treatment in human biliary tract cancer cell lines." Photochem Photobiol Sci **6**(6): 619-627.
Kim, D. H., S. Y. Kim, et al. (2011). "Assessment of Chemotherapy Response Using FDG-PET in Pediatric Bone Tumors: A Single Institution Experience." Cancer Res Treat **43**(3): 170-175.
Kim, S. Y., C. H. Lee, et al. (2008). "Inhibition of the CXCR4/CXCL12 chemokine pathway reduces the development of murine pulmonary metastases." Clin Exp Metastasis **25**(3): 201-211.
Klein, M. J. and G. P. Siegal (2006). "Osteosarcoma: anatomic and histologic variants." Am J Clin Pathol **125**(4): 555-581.
Knuuti, J. and F. M. Bengel (2008). "Positron emission tomography and molecular imaging." Heart **94**(3): 360-367.
Korbelik, M., G. Krosl, et al. (1996). "The role of host lymphoid populations in the response of mouse EMT6 tumor to photodynamic therapy." Cancer Res **56**(24): 5647-5652.
Korbelik, M., J. Sun, et al. (2001). "Interaction between photodynamic therapy and BCG immunotherapy responsible for the reduced recurrence of treated mouse tumors." Photochem Photobiol **73**(4): 403-409.
Kousis, P. C., B. W. Henderson, et al. (2007). "Photodynamic therapy enhancement of antitumor immunity is regulated by neutrophils." Cancer Res **67**(21): 10501-10510.
Kurohane, K., A. Tominaga, et al. (2001). "Photodynamic therapy targeted to tumor-induced angiogenic vessels." Cancer Lett **167**(1): 49-56.
Kusuzaki, K., G. Minami, et al. (2000). "Photodynamic inactivation with acridine orange on a multidrug-resistant mouse osteosarcoma cell line." Jpn J Cancer Res **91**(4): 439-445.
Kwee, S. A., T. R. DeGrado, et al. (2007). "Cancer imaging with fluorine-18-labeled choline derivatives." Semin Nucl Med **37**(6): 420-429.
Kwee, S. A., J. P. Ko, et al. (2007). "Solitary brain lesions enhancing at MR imaging: evaluation with fluorine 18 fluorocholine PET." Radiology **244**(2): 557-565.
Lamoureux, F., V. Trichet, et al. (2007). "Recent advances in the management of osteosarcoma and forthcoming therapeutic strategies." Expert Rev Anticancer Ther **7**(2): 169-181.

Langen, K. J., K. Hamacher, et al. (2006). "O-(2-[18F]fluoroethyl)-L-tyrosine: uptake mechanisms and clinical applications." Nucl Med Biol 33(3): 287-294.

Langsteger, W., M. Heinisch, et al. (2006). "The role of fluorodeoxyglucose, 18F-dihydroxyphenylalanine, 18F-choline, and 18F-fluoride in bone imaging with emphasis on prostate and breast." Semin Nucl Med 36(1): 73-92.

Larson, S. M., M. Morris, et al. (2004). "Tumor localization of 16beta-18F-fluoro-5alpha-dihydrotestosterone versus 18F-FDG in patients with progressive, metastatic prostate cancer." J Nucl Med 45(3): 366-373.

Lassalle, H. P., D. Dumas, et al. (2009). "Correlation between in vivo pharmacokinetics, intratumoral distribution and photodynamic efficiency of liposomal mTHPC." J Control Release 134(2): 118-124.

Lassalle, H. P., M. Wagner, et al. (2008). "Fluorescence imaging of Foscan and Foslip in the plasma membrane and in whole cells." J Photochem Photobiol B 92(1): 47-53.

Lee, J. A., J. S. Jung, et al. (2011). "RANKL expression is related to treatment outcome of patients with localized, high-grade osteosarcoma." Pediatr Blood Cancer 56(5): 738-743.

Leung, W. N., X. Sun, et al. (2002). "Photodynamic effects of mTHPC on human colon adenocarcinoma cells: photocytotoxicity, subcellular localization and apoptosis." Photochem Photobiol 75(4): 406-411.

Li, F., S. Sone, et al. (2004). "Malignant versus benign nodules at CT screening for lung cancer: comparison of thin-section CT findings." Radiology 233(3): 793-798.

Lindholm, P., H. Minn, et al. (1993). "Influence of the blood glucose concentration on FDG uptake in cancer--a PET study." J Nucl Med 34(1): 1-6.

Litle, V. R., N. A. Christie, et al. (2003). "Photodynamic therapy for endobronchial metastases from nonbronchogenic primaries." Ann Thorac Surg 76(2): 370-375; discussion 375.

Longhi, A., C. Errani, et al. (2006). "Primary bone osteosarcoma in the pediatric age: state of the art." Cancer Treat Rev 32(6): 423-436.

Luksiene, Z. (2003). "Photodynamic therapy: mechanism of action and ways to improve the efficiency of treatment." Medicina (Kaunas) 39(12): 1137-1150.

Manara, M. C., N. Baldini, et al. (2000). "Reversal of malignant phenotype in human osteosarcoma cells transduced with the alkaline phosphatase gene." Bone 26(3): 215-220.

Manara, M. C., M. Serra, et al. (2004). "Effectiveness of Type I interferons in the treatment of multidrug resistant osteosarcoma cells." Int J Oncol 24(2): 365-372.

Manic, S., L. Gatti, et al. (2003). "Mechanisms controlling sensitivity to platinum complexes: role of p53 and DNA mismatch repair." Curr Cancer Drug Targets 3(1): 21-29.

Mareel, M., M. J. Oliveira, et al. (2009). "Cancer invasion and metastasis: interacting ecosystems." Virchows Arch 454(6): 599-622.

Mathot, L. and J. Stenninger (2011). "Behavior of seeds and soil in the mechanism of metastasis: A deeper understanding." Cancer Sci.

McCarville, M. B., H. M. Lederman, et al. (2006). "Distinguishing benign from malignant pulmonary nodules with helical chest CT in children with malignant solid tumors." Radiology 239(2): 514-520.

Mehlen, P. and A. Puisieux (2006). "Metastasis: a question of life or death." Nat Rev Cancer 6(6): 449-458.

Mertens, K., D. Slaets, et al. (2010). "PET with (18)F-labelled choline-based tracers for tumour imaging: a review of the literature." Eur J Nucl Med Mol Imaging 37(11): 2188-2193.

Messa, C., C. Landoni, et al. (2000). "Is there a role for FDG PET in the diagnosis of musculoskeletal neoplasms?" J Nucl Med 41(10): 1702-1703.

Meyer, W. H. (1991). "Recent developments in genetic mechanisms, assessment, and treatment of osteosarcomas." Curr Opin Oncol 3(4): 689-693.

Meyers, P. A., G. Heller, et al. (1992). "Chemotherapy for nonmetastatic osteogenic sarcoma: the Memorial Sloan-Kettering experience." J Clin Oncol 10(1): 5-15.

Mikolajczyk, K., M. Szabatin, et al. (1998). "A JAVA environment for medical image data analysis: initial application for brain PET quantitation." Med Inform (Lond) 23(3): 207-214.

Mintun, M. A., M. J. Welch, et al. (1988). "Breast cancer: PET imaging of estrogen receptors." Radiology 169(1): 45-48.

Mirabello, L., R. J. Troisi, et al. (2009). "International osteosarcoma incidence patterns in children and adolescents, middle ages and elderly persons." Int J Cancer **125**(1): 229-234.

Mirabello, L., R. J. Troisi, et al. (2009). "Osteosarcoma incidence and survival rates from 1973 to 2004: data from the Surveillance, Epidemiology, and End Results Program." Cancer **115**(7): 1531-1543.

Molthoff, C. F., B. M. Klabbers, et al. (2007). "Monitoring response to radiotherapy in human squamous cell cancer bearing nude mice: comparison of 2'-deoxy-2'-[18F]fluoro-D-glucose (FDG) and 3'-[18F]fluoro-3'-deoxythymidine (FLT)." Mol Imaging Biol **9**(6): 340-347.

Muller, C. R., S. Smeland, et al. (2005). "Interferon-alpha as the only adjuvant treatment in high-grade osteosarcoma: long term results of the Karolinska Hospital series." Acta Oncol **44**(5): 475-480.

Mumprecht, V., M. Honer, et al. (2010). "In vivo imaging of inflammation- and tumor-induced lymph node lymphangiogenesis by immuno-positron emission tomography." Cancer Res **70**(21): 8842-8851.

Mundy, G. R. (2002). "Metastasis to bone: causes, consequences and therapeutic opportunities." Nat Rev Cancer **2**(8): 584-593.

Myers, R. C., B. H. Lau, et al. (1989). "Modulation of hematoporphyrin derivative-sensitized phototherapy with corynebacterium parvum in murine transitional cell carcinoma." Urology **33**(3): 230-235.

Nguyen, D. X., P. D. Bos, et al. (2009). "Metastasis: from dissemination to organ-specific colonization." Nat Rev Cancer **9**(4): 274-284.

Nguyen, D. X. and J. Massague (2007). "Genetic determinants of cancer metastasis." Nat Rev Genet **8**(5): 341-352.

Nomura, J., S. Yanase, et al. (2004). "Efficacy of combined photodynamic and hyperthermic therapy with a new light source in an in vivo osteosarcoma tumor model." J Clin Laser Med Surg **22**(1): 3-8.

Nowell, P. C. (1976). "The clonal evolution of tumor cell populations." Science **194**(4260): 23-28.

O'Connor, A. E., W. M. Gallagher, et al. (2009). "Porphyrin and nonporphyrin photosensitizers in oncology: preclinical and clinical advances in photodynamic therapy." Photochem Photobiol **85**(5): 1053-1074.

Oleinick, N. L., R. L. Morris, et al. (2002). "The role of apoptosis in response to photodynamic therapy: what, where, why, and how." Photochem Photobiol Sci **1**(1): 1-21.

Ost, D. and A. Fein (2004). "Management strategies for the solitary pulmonary nodule." Curr Opin Pulm Med **10**(4): 272-278.

Ost, D., A. M. Fein, et al. (2003). "Clinical practice. The solitary pulmonary nodule." N Engl J Med **348**(25): 2535-2542.

Ottaviani, G. and N. Jaffe (2009). "The epidemiology of osteosarcoma." Cancer Treat Res **152**: 3-13.

Padua, D., X. H. Zhang, et al. (2008). "TGFbeta primes breast tumors for lung metastasis seeding through angiopoietin-like 4." Cell **133**(1): 66-77.

Pelosi, E., V. Arena, et al. (2008). "Role of whole-body 18F-choline PET/CT in disease detection in patients with biochemical relapse after radical treatment for prostate cancer." Radiol Med **113**(6): 895-904.

Penuelas, I., G. Mazzolini, et al. (2005). "Positron emission tomography imaging of adenoviral-mediated transgene expression in liver cancer patients." Gastroenterology **128**(7): 1787-1795.

Perissinotto, E., G. Cavalloni, et al. (2005). "Involvement of chemokine receptor 4/stromal cell-derived factor 1 system during osteosarcoma tumor progression." Clin Cancer Res **11**(2 Pt 1): 490-497.

Peterson, J. J. (2007). "F-18 FDG-PET for detection of osseous metastatic disease and staging, restaging, and monitoring response to therapy of musculoskeletal tumors." Semin Musculoskelet Radiol **11**(3): 246-260.

Phelps, M. E. (2000). "Positron emission tomography provides molecular imaging of biological processes." Proc Natl Acad Sci U S A **97**(16): 9226-9233.

Phelps, M. E., E. J. Hoffman, et al. (1975). "Application of annihilation coincidence detection to transaxial reconstruction tomography." J Nucl Med **16**(3): 210-224.

Picci, P. (2007). "Osteosarcoma (osteogenic sarcoma)." Orphanet J Rare Dis **2**: 6.

Potter, C. and A. L. Harris (2004). "Hypoxia inducible carbonic anhydrase IX, marker of tumour hypoxia, survival pathway and therapy target." Cell Cycle 3(2): 164-167.

Price, C. H. (1955). "Osteogenic sarcoma; an analysis of the age and sex incidence." Br J Cancer 9(4): 558-574.

Rajendran, J. G., D. L. Schwartz, et al. (2006). "Tumor hypoxia imaging with [F-18] fluoromisonidazole positron emission tomography in head and neck cancer." Clin Cancer Res 12(18): 5435-5441.

Raman, V., D. Artemov, et al. (2006). "Characterizing vascular parameters in hypoxic regions: a combined magnetic resonance and optical imaging study of a human prostate cancer model." Cancer Res 66(20): 9929-9936.

Reidy, K., C. Campanile, et al. (2012). "mTHPC-Mediated Photodynamic Therapy is Effective in the Metastatic Human 143B Osteosarcoma Cells." Photochem Photobiol 88(3): 721-727.

Reiners, J. J., Jr., P. Agostinis, et al. (2010). "Assessing autophagy in the context of photodynamic therapy." Autophagy 6(1): 7-18.

Reske, S. N. and S. Deisenhofer (2006). "Is 3'-deoxy-3'-(18)F-fluorothymidine a better marker for tumour response than (18)F-fluorodeoxyglucose?" Eur J Nucl Med Mol Imaging 33 Suppl 1: 38-43.

Rischin, D., R. J. Hicks, et al. (2006). "Prognostic significance of [18F]-misonidazole positron emission tomography-detected tumor hypoxia in patients with advanced head and neck cancer randomly assigned to chemoradiation with or without tirapazamine: a substudy of Trans-Tasman Radiation Oncology Group Study 98.02." J Clin Oncol 24(13): 2098-2104.

Ritter, J. and S. S. Bielack (2010). "Osteosarcoma." Ann Oncol 21 Suppl 7: vii320-325.

Sabile, A. A., M. J. Arlt, et al. (2011). "Cyr61 expression in Osteosarcoma indicates poor prognosis and promotes intratibial growth and lung metastasis in mice." J Bone Miner Res.

Sabile, A. A., M. J. Arlt, et al. (2013). "Caprin-1, a novel Cyr61-interacting protein, promotes osteosarcoma tumor growth and lung metastasis in mice." Biochim Biophys Acta 1832(8): 1173-1182.

Saji, H., W. Song, et al. (2006). "Systemic antitumor effect of intratumoral injection of dendritic cells in combination with local photodynamic therapy." Clin Cancer Res 12(8): 2568-2574.

Saleem, A., J. Yap, et al. (2000). "Modulation of fluorouracil tissue pharmacokinetics by eniluracil: in-vivo imaging of drug action." Lancet 355(9221): 2125-2131.

Savage, S. A. and L. Mirabello (2011). "Using epidemiology and genomics to understand osteosarcoma etiology." Sarcoma 2011: 548151.

Schlyer, D. J. (2004). "PET tracers and radiochemistry." Ann Acad Med Singapore 33(2): 146-154.

Schrager, J., R. E. Patzer, et al. (2011). "Survival outcomes of pediatric osteosarcoma and Ewing's sarcoma: a comparison of surgery type within the SEER database, 1988-2007." J Registry Manag 38(3): 153-161.

Schuetze, S. M., B. P. Rubin, et al. (2005). "Use of positron emission tomography in localized extremity soft tissue sarcoma treated with neoadjuvant chemotherapy." Cancer 103(2): 339-348.

Scully, S. P., M. A. Ghert, et al. (2002). "Pathologic fracture in osteosarcoma : prognostic importance and treatment implications." J Bone Joint Surg Am 84-A(1): 49-57.

Serra, M., K. Scotlandi, et al. (2003). "Value of P-glycoprotein and clinicopathologic factors as the basis for new treatment strategies in high-grade osteosarcoma of the extremities." J Clin Oncol 21(3): 536-542.

Shani, J. and W. Wolf (1977). "A model for prediction of chemotherapy response to 5-fluorouracil based on the differential distribution of 5-[18F]fluorouracil in sensitive versus resistant lymphocytic leukemia in mice." Cancer Res 37(7 Pt 1): 2306-2308.

Silva, C. T., J. G. Amaral, et al. (2010). "CT characteristics of lung nodules present at diagnosis of extrapulmonary malignancy in children." AJR Am J Roentgenol 194(3): 772-778.

Snyder, J. W., W. R. Greco, et al. (2003). "Photodynamic therapy: a means to enhanced drug delivery to tumors." Cancer Res 63(23): 8126-8131.

Spaeth, N., M. T. Wyss, et al. (2004). "Uptake of 18F-fluorocholine, 18F-fluoroethyl-L-tyrosine, and 18F-FDG in acute cerebral radiation injury in the rat: implications for separation of radiation necrosis from tumor recurrence." J Nucl Med 45(11): 1931-1938.

Steeg, P. S. (2003). "Metastasis suppressors alter the signal transduction of cancer cells." Nat Rev Cancer **3**(1): 55-63.
Stein, U., W. Walther, et al. (2009). "MACC1, a newly identified key regulator of HGF-MET signaling, predicts colon cancer metastasis." Nat Med **15**(1): 59-67.
Stiller, C. A., S. S. Bielack, et al. (2006). "Bone tumours in European children and adolescents, 1978-1997. Report from the Automated Childhood Cancer Information System project." Eur J Cancer **42**(13): 2124-2135.
Strander, H. and S. Einhorn (1977). "Effect of human leukocyte interferon on the growth of human osteosarcoma cells in tissue culture." Int J Cancer **19**(4): 468-473.
Stroobants, S., J. Goeminne, et al. (2003). "18FDG-Positron emission tomography for the early prediction of response in advanced soft tissue sarcoma treated with imatinib mesylate (Glivec)." Eur J Cancer **39**(14): 2012-2020.
Sugiyama, M., H. Sakahara, et al. (2004). "Evaluation of 3'-deoxy-3'-18F-fluorothymidine for monitoring tumor response to radiotherapy and photodynamic therapy in mice." J Nucl Med **45**(10): 1754-1758.
Sun, Y. X., J. Wang, et al. (2003). "Expression of CXCR4 and CXCL12 (SDF-1) in human prostate cancers (PCa) in vivo." J Cell Biochem **89**(3): 462-473.
Sundaram, M. (1997). "The use of gadolinium in the MR imaging of bone tumors." Semin Ultrasound CT MR **18**(4): 307-311.
Swinnen, J. V., K. Brusselmans, et al. (2006). "Increased lipogenesis in cancer cells: new players, novel targets." Curr Opin Clin Nutr Metab Care **9**(4): 358-365.
Talbot, J. N., F. Gutman, et al. (2006). "PET/CT in patients with hepatocellular carcinoma using [(18)F]fluorocholine: preliminary comparison with [(18)F]FDG PET/CT." Eur J Nucl Med Mol Imaging **33**(11): 1285-1289.
Teiten, M. H., L. Bezdetnaya, et al. (2003). "Endoplasmic reticulum and Golgi apparatus are the preferential sites of Foscan localisation in cultured tumour cells." Br J Cancer **88**(1): 146-152.
Thie, J. A. (2004). "Understanding the standardized uptake value, its methods, and implications for usage." J Nucl Med **45**(9): 1431-1434.
Torizuka, T., K. R. Zasadny, et al. (1999). "Diabetes Decreases FDG Accumulation in Primary Lung Cancer." Clin Positron Imaging **2**(5): 281-287.
Tsang, W. P., S. P. Chau, et al. (2003). "Modulation of multidrug resistance-associated protein 1 (MRP1) by p53 mutant in Saos-2 cells." Cancer Chemother Pharmacol **51**(2): 161-166.
Uehara, M., K. Sano, et al. (2000). "Enhancement of the photodynamic antitumor effect by streptococcal preparation OK-432 in the mouse carcinoma." Cancer Immunol Immunother **49**(8): 401-409.
Weber, W. A. (2006). "Positron emission tomography as an imaging biomarker." J Clin Oncol **24**(20): 3282-3292.
Weber, W. A., K. Ott, et al. (2001). "Prediction of response to preoperative chemotherapy in adenocarcinomas of the esophagogastric junction by metabolic imaging." J Clin Oncol **19**(12): 3058-3065.
Wester, H. J., M. Herz, et al. (1999). "Synthesis and radiopharmacology of O-(2-[18F]fluoroethyl)-L-tyrosine for tumor imaging." J Nucl Med **40**(1): 205-212.
Whelan, J., D. Patterson, et al. (2010). "The role of interferons in the treatment of osteosarcoma." Pediatr Blood Cancer **54**(3): 350-354.
Widhe, B. and T. Widhe (2000). "Initial symptoms and clinical features in osteosarcoma and Ewing sarcoma." J Bone Joint Surg Am **82**(5): 667-674.
Wilson, W. R. and M. P. Hay (2011). "Targeting hypoxia in cancer therapy." Nat Rev Cancer **11**(6): 393-410.
Wirth, C. and W. Winkelmann (2004). Orthopädie und Orthopädische Chirurgie Tumoren und tumourännliche Erkrankungen. Stuttgart, Thieme.
Wittig, J. C., J. Bickels, et al. (2002). "Osteosarcoma of the proximal humerus: long-term results with limb-sparing surgery." Clin Orthop Relat Res(397): 156-176.
Yamamoto, Y., Y. Nishiyama, et al. (2007). "Correlation of 18F-FLT and 18F-FDG uptake on PET with Ki-67 immunohistochemistry in non-small cell lung cancer." Eur J Nucl Med Mol Imaging **34**(10): 1610-1616.

Yang, J. and R. A. Weinberg (2008). "Epithelial-mesenchymal transition: at the crossroads of development and tumor metastasis." Dev Cell **14**(6): 818-829.

Yang, Y. J., J. S. Ryu, et al. (2006). "Use of 3'-deoxy-3'-[18F]fluorothymidine PET to monitor early responses to radiation therapy in murine SCCVII tumors." Eur J Nucl Med Mol Imaging **33**(4): 412-419.

Yarmish, G., M. J. Klein, et al. (2010). "Imaging characteristics of primary osteosarcoma: nonconventional subtypes." Radiographics **30**(6): 1653-1672.

Yaw, K. M. (1999). "Pediatric bone tumors." Semin Surg Oncol **16**(2): 173-183.

Yin, J. J., K. Selander, et al. (1999). "TGF-beta signaling blockade inhibits PTHrP secretion by breast cancer cells and bone metastases development." J Clin Invest **103**(2): 197-206.

Yow, C. M., J. Y. Chen, et al. (2000). "Cellular uptake, subcellular localization and photodamaging effect of temoporfin (mTHPC) in nasopharyngeal carcinoma cells: comparison with hematoporphyrin derivative." Cancer Lett **157**(2): 123-131.

Yuan, J., C. Ossendorf, et al. (2009). "Osteoblastic and osteolytic human osteosarcomas can be studied with a new xenograft mouse model producing spontaneous metastases." Cancer Invest **27**(4): 435-442.

Zimmermann, A., M. Ritsch-Marte, et al. (2001). "mTHPC-mediated photodynamic diagnosis of malignant brain tumors." Photochem Photobiol **74**(4): 611-616.

7 ACKNOWLEDGMENTS

The work presented here was performed at the Department of Orthopedic Research at the University Clinic Balgrist in Zurich. A significant part of the experiments concerning Positron Emission Tomography was also performed at the Animal Imaging Center located in the Institute of Pharmaceutical Sciences (ETH-Zurich). Therefore the people to be acknowledged for their enormous help will be quite long.

First of all I would like to thank Prof. Dr. Roger Schibli that accepted to be my thesis director and that guided me to carry on this multidisciplinary thesis and that was always supporting me with informative advices.

My deepest gratitude goes to Prof. Dr Bruno Fuchs and Prof. Dr. Walter Born that welcomed me in their laboratory always open for scientific discussions, for bright suggestions and for the support throughout the four years.

I would furthermore like to thank Prof. Dr. Simon Mensah Ametamey and Prof. Dr. Michael Detmar to accept being part of my Thesis Committee and for their critical and helpful contribution during the yearly Committee Meetings.

A big thank you goes to my collaborators at ETH-Hönggerberg that helped technically with all the work that I have performed there. In particular I wanted to express my gratitude to Claudia Keller and Petra Wirth that stayed with me long days in the Animal Imaging Center for the imaging; Dr. Michael Honer that introduced me to the magic world of the pMod software; PD Dr. Stefanie Krämer for the final check of the images analysis and Cindy Fischer and Mathias Nobst for getting up even before sunrise to produce the PET tracers for me.

i want morebooks!

Buy your books fast and straightforward online - at one of world's fastest growing online book stores! Environmentally sound due to Print-on-Demand technologies.

Buy your books online at

www.get-morebooks.com

Kaufen Sie Ihre Bücher schnell und unkompliziert online – auf einer der am schnellsten wachsenden Buchhandelsplattformen weltweit! Dank Print-On-Demand umwelt- und ressourcenschonend produziert.

Bücher schneller online kaufen

www.morebooks.de

VDM Verlagsservicegesellschaft mbH
Heinrich-Böcking-Str. 6-8
D - 66121 Saarbrücken

Telefon: +49 681 3720 174
Telefax: +49 681 3720 1749

info@vdm-vsg.de
www.vdm-vsg.de

Printed by Books on Demand GmbH, Norderstedt / Germany